D1676857

THE BACTERIA: THEIR ORIGIN, STRUCTURE, FUNCTION AND ANTIBIOSIS

The Bacteria: Their Origin, Structure, Function and Antibiosis

by

Arthur L. Koch

Indiana University,
Bloomington, IN, USA

A C.I.P. Catalogue record for this book is available from the Library of Congress.

ISBN-10 1-4020-3205-6 (HB)
ISBN-13 978-1-4020-3205-9 (HB)
ISBN-10 1-4020-3206-4 (e-book)
ISBN-13 978-1-4020-3206-6 (e-book)

Published by Springer,
P.O. Box 17, 3300 AA Dordrecht, The Netherlands.

www.springer.com

Printed on acid-free paper

Printed in the Netherlands.

Table of Contents

Preface

This book may seem like three or four books even though the main focus is on a specialized topic—the bacterial cell wall. Its job is to formulate the innovations that caused life to initiate on earth, those that caused cell physiology to develop without diversity developing, those that allowed the murein walls of the cells to arise, those that led to the separation of the domain of Bacteria from other organisms, those that allowed the Archaea and the Eukarya to develop independently, and those that then led to the development of a very diverse biosphere. It must have taken a long time after the origin of the first cell; evolution had proceeded to produce very effective organisms. At some point a collection of very similar organisms arose that were first called collectively the Last Universal Ancestor (LUA), and stable divergence developed from there. The first bacterium had a protective cell wall and its descendants developed in many diverse evolutionary directions, gave rise to many species of bacteria with various life strategies, and expanded to fill the many niches in the collection.

As the kingdoms or domains of Archaea and Eukarya evolved, many of these organisms (and even some bacteria) acted against bacteria. The development of antibiotics acting on the wall of bacteria and lytic enzymes, called lysozymes, produced by protozoa, plants and animals, led to destruction of many bacteria. These antagonistic challenges to bacteria resulted from its own cell wall structure. This structure was both bacteria's most prominent advantage and its greatest liability. It led to growth success of bacteria and to development of a widespread domain—and to their destruction by other organisms. Man subsequently extended and elaborated these destructive tricks against bacteria, which led to the antibiotic era of medicine. Sometimes however, medical progress has turned out to be retrograde to long term medical advances. The attempt here is present the physics, chemistry and the evolution of life forms that created targets for antibiotics and the bacterial response to antibiotics. Logically all aspects must be considered together in order that new treatments for infection will not only work but will be effective for a long time.

Most of today's bacteria maintain a peptidoglycan or murein wall (called a sacculus or exoskeleton) that surrounds them completely. This strong wall protects them against osmotic differences between the inside and the outside that otherwise might lead to the influx of water and the resultant rupture of the cell, but it also has many roles in the biology of bacteria. This book focuses first on the chemistry and mechanics of the cell's wall formation and function and how evolutionary forces probably led to its development. The makeup and structure of the wall permit bacteria to occupy diverse habitats and niches. It allows bacteria to do the same things that larger, multicellular organisms do, but in different ways. The book questions and tries to answer: (1) How does the bacterial envelope enlarge safely with the maintenance of cell shape? (2) How does the wall function as a critical cell organelle for other vital bacterial needs? (3) How does division of the sacculus and the bacterial cell occur? (4) How do other living organisms (the Archaea, Eukarya, and particularly, *Homo sapiens*) combat bacteria? (5) How have bacteria evolved in the recent past to overcome human production and distribution of wall-directed antibiotics? The ideas presented are logical conclusions from what we do know, but only tentative answers can be given because ideas about early evolution and more recent evolution have been deduced largely from properties of modern organisms and the current knowledge of molecular genomics.

Arguments throughout the book are mustered to illustrate that prokaryotic life is more directly and simply dependent on physical and chemical principles than are the life forms of multicellular organisms. Of course, either directly or indirectly, the exploitation of physical laws and chemical reactions is dependent on Darwin's three principles. In modern translation these are (1) replication of informational biomolecules must be accurate most of the time; (2) only occasionally do mutations take place, and (3) the translation into functional working forms takes place from the information propagated in molecular form. The role of antibiosis in evolution and man's attempt to use it is for his own advantage. This includes the successes and the failures. In the future, the concepts presented in this book, I am certain, will be critical for medical advances in treatment of infectious diseases.

Legend to the Frontispiece:

The Structural Elements of the Bacterial Wall: Five Disaccharide Penta-muropeptides Forming a Glycan Chain and Two about to Form a Nona-muropeptide

Seven penta-muropeptides of *Escherichia coli* or *Bacillus subtilis* are shown in this frontispiece. The top portion shows a glycan chain formed from five NAG (N-Acetyl-Glucosamine) and five alternating NAM (N-Acetyl-Muramic acid) residues. The NAM residues are each linked with penta-muropeptides. The conformation of the glycan chain is a spiral with the top and bottom muropeptides in the same plane. The middle one is also in the same plane but points in the opposite direction. Two of these structures in mirror images are bound to each other by transpeptidation with the loss of the terminal D-Alanine (D-Ala) groups that are shown enclosed in blue ovals. This forms a tessera as shown more explicitly in Figs. 8.1 and 8.2. The bacterial sacculus is formed of many such tesserae and completely encloses the cell.

In the formation of a crossbridge the D-Alanine would be removed in the transpeptidation process and resulting disaccharide tetra-muropeptide coupled by endopeptide bonds (tail-to-tail bonds) with another disaccharide penta-muropeptide to form nine-membered nona-muropeptide. Other muropeptides are shown in the top part of the figure, two pointing above and two pointing below the plane of the cell; they either remain as such or are degraded. The remaining muropeptide is in the plane but points in the opposite direction.

In the bottom part of the figure, the terminal D-Ala groups are again shown in blue and the zwitter ionic groups of the diaminopimelic acid groups are indicated within yellow ellipses. They are placed correctly, but the bonding is not shown because they are part of the diaminopimelic acid group. This amino acid is abbreviated A_2pm (and also DAP in the literature) Diaminopimelic acid minus the dipeptide D-Ala-D-Ala and also minus the zwitter group is designated for clarity by ZZ. Two penta-muropeptides are shown positioned for the removal of the terminal D-Ala and formation of the tail-to-tail bond with the amino group of the zwitter ion. The two ways that this can be done are indicated. One precludes the other, but the remaining unbound D-Ala-D-Ala and the zwitter of the diaminopimelic acid have ionic attraction to each other and this sterically precludes entry of additional muropeptides or the endo-transpeptidase enzyme as the nona-muropeptide molecule is originally formed. However, when the formed nona-muropeptide is stressed enough by growth to break the ionic attachment of these two groups, then entry becomes allowed and further growth is possible.

Part 1

Origin of Bacteria

Chapter 1

The Origin of Life Based on Physical Principles

All the life forms known to us are completely dependent on organic and physical chemistry. Life depends on the chemistry of carbon compounds and on the laws concerning chemical reactions. Life depends on a temperature low enough to allow covalent bonds to form, with enzyme assistance, but generally not at a high enough temperature to break chemical bonds. Life depends on having an aqueous environment within cells where vital reactions can take place. Living machines function as isothermal engines that require the abstraction of available energy from the environment into special forms that a cell can use. The energy is used to drive chemical reactions in non-favorable directions in order to form molecular and supramolecular structures necessary for cellular life to take place. In this chapter, the essential thermodynamics of life are stated and the real (anthropomorphic) meaning of "free energy" elucidated. Free energy is the basis of biochemistry but is not fully explained in modern courses in chemistry and biology nor, unfortunately, in microbiology.

THOUGHTS ABOUT THE ORIGIN OF LIFE

For a student of molecular biology, the first logical thought about the origin of life certainly would be that, since the system of specific enzymes and the ways to make them are so complicated and their number so extensive, life could not have arisen spontaneously in any imaginable physical and chemical environment no matter how long a time period was involved. There are two extreme rationalizations for the existence of life from this thought: either Divine Creation took place or repetitive random chemical reactions occurred until some combination worked to generate a growing and reproducing cell. Neither extreme seems reasonable to me. However, if many thousands of enzymes are needed to function to activate a reproducing cell, then nothing between the two is logical. On the Divine side, the universe obeys certain rules. Although the origin of these laws is unclear, the laws themselves are becoming clear, at least for our universe. Could they have been different than they are? If so, there is a role for God under any name. If by chance the Universal Laws had been different in different universes, then maybe we are just lucky to have been in this particular one. At the other extreme, given whatever rules might apply to other universes, a very large number of trials might be needed in most in order to generate a "First Cell". In any case, it may have been necessary to once get a First Cell that lives, grows, and reproduces (Koch, 1985; Koch, 2001; Koch and Silver, 2005, and see the references in Koch and Silver) for life to have persisted on earth. It could subsequently grow, evolve, mutate, and differentiate. Life, of course, must initially have been started by chance with all the necessary paraphernalia to make a minimal, but working and reproducing cell.

Between these two extremes, and with the blessing of Darwin, we can imagine that life started very simply but was hobbled by the many things it could not do. The First Cell was probably very minimally endowed. However, long-term evolution, after the primeval First Cell arose, slowly generated the large number of capabilities possessed by all modern cells and subsequently many kinds of organisms. The evolutionary process that created cells with the essential cellular physiological capability possessed by modern cells must have taken a very long time (I guess almost a billion years). The generation of the diversity of organisms, their organization into multicellular organisms, and their interactions with each other occurred later. In the latter part of this book the interaction between bacteria and people and antibiotics will be stressed. However, this is simply another consequence of the development of a complex biology on earth.

First we need to consider how a chemical machine can work. The structure of any man-made machine allows an action on one part of it to cause a specific action on another of its parts. The machine's construction, or the repetitive construction to make many copies of a machine, requires an external entity: man. On the other hand, while an outside agent does construct each living machine similar to the way an assembly line constructs automobiles, the basis of life, at least in this universe, is that spontaneously the elements of chemistry, those of physical chemistry, the three laws of thermodynamics, available resources, and the subsequent occurrence of a lucky vesicle that was the ancestor of us all, caused growth and reproduction of living systems to occur.

ASTROBIOLOGY

This is not the place for discussion of the development of stars, planets, or the chemistry and environments conducive to life and energy transduction. These must have been involved in development of living systems. Recent reviews can be found in Brack (1988) and Koch and Silver (2005).

THREE LAWS OF THERMODYNAMICS

Various creationists make much of the fact that according to the Second Law of Thermodynamics, any system will gradually and eventually achieve its most stable equilibrium state. This state is, evidently, a non-living state. The creationist's thermodynamics is entirely correct. For example, if the earth were a closed system, it could not support life. When the sun grows cold, everything on earth will then become lifeless. There is no doubt that this will happen. The point is that our earth, for now, is not a closed but a dynamic system because radiant energy falls on it and infrared radiation leaves the earth; thus it has the hallmarks of an open system. However, the thermodynamic rules do permit some of the radiant energy to be trapped (at least temporarily), and it is this tidbit of energy that allows life to exist. The sun plus planets plus other intra-planetary materials out to some very large distance from the sun would approximate a closed system and one that slowly would be "winding down".

The laws of thermodynamics will never be proved, but, also, they will never be disproved. They are statements of "impotence"; they state only what cannot happen. At least, these laws have not yet been observed to have been violated. The First Law, "The conservation of energy", states that although you can convert energy from one form to another, the amount of energy will neither be increased nor decreased. The Second Law states that work cannot be converted totally from thermal energy to mechanical or electrical energy by any machine, except under a special condition stated by the Third Law. The Third Law is that, for a thermal machine, only if the reservoir to which the heat is delivered is maintained at absolute zero can the thermal energy be converted completely into other kinds of energy. Does this matter to a biological system? After all, life is isothermal; it should therefore function with zero efficiency as a heat engine. The answer is that it is not a heat engine. Except for photosynthesis, life is a chemical engine acting by converting covalent bonds.

The laws of thermodynamics are fundamental to life. A humorous paraphrase of the three laws of thermodynamics is (First Law) you can't get something for nothing, (Second Law) there is tax on it, and (Third Law) only at absolute zero is there no tax. When stated less frivolously, these laws are the basis of how all living cells function through chemistry and without physical connection between the parts, such as cogs, wheels, and drive shafts, although these things are necessary in an automobile's function or in any other mechanical machine. To state the laws in yet another way, the First Law states that (except for thermal energy) energy can be converted from one form to another without gain or loss. The Second Law states that a heat engine's mechanical efficiency depends on the temperatures of the source and sink. The Third Law states that thermal energy can be converted to another energy form with an efficiency of 100% only if the sink is at absolute zero. This leads to the implication that a biological machine would have 0% efficiency because it is isothermal. This is wrong because it is not a heat engine, instead it is a chemical engine and quite different rules apply.

The way the cellular machine works is by having catalysts that favor only certain chemical reactions. This ability to catalyze certain specific reactions and not others is the major virtue of the living machine. Stated more strongly, the specificity of an enzyme is such that it does not catalyze other reactions than the one it was "designed" for. Add to this a little bit of structure and life could work. That structure was initially abiotic (in the first living generation, for example, life was a vesicle probably created by environmental wave action). In addition, there has been extensive evolution of cell structures after the First Cell. These include: chromosomes instead of unlinked genes, cell walls instead of no walls, protein aggregates like muscle fibrils to do mechanical work and cytoskeletons to support the cellular structure. Add the existence of binding proteins and bound carbohydrates on the outsides of cells allowing them to interact and add other cells, and with little bit more development, here we are.

CHEMICAL REACTIONS

At too high a temperature, say in excess of 2000 °F, carbon, hydrogen, oxygen, nitrogen, sulfur, and phosphorus atoms will not form stable covalent bonds. Neither will they at very low temperatures. In the latter situation they may nearly come together but will not bind to each other because they do not have enough "energy of activation" and, subsequently will come apart and not be bound. In the former case the atoms may be bound temporarily to each other, but will have enough energy of activation to break these bonds and come apart. Life based on organic chemistry, as a result, is limited to a very narrow temperature range. The range is further limited because life depends on favoring only reactions useful to a cell. This dependency on catalysts (enzymes) to speed up some reactions is vital. Note that the key point is that they do not catalyze many other reactions. In the living world today, the cellular edifice has many thousands of vital enzymes to chart the course that it pursues. Nothing less would be sufficient in our competitive world.

THE MECHANICS OF COVALENT BONDING

At very high temperatures, as found in stars, there are nucleons but no atoms. At lower temperatures there will be the formation of stable atoms because the temperature does not dissociate the parts of atoms. At the lower temperature the attraction of positive and negative charges overcomes the thermal repulsion. At still lower temperature the atoms will have equal numbers of positive and negative charges. At even lower temperatures a more stable state results when the positive charges of two atomic nuclei attract, because of associated negatively charged electrons, and they share electrons. This results in a force that holds the atoms together and effectively creates "the bond". This covalent bond is the glue that makes organic structures stable. This is not to say that weaker bonds are irrelevant. They are very relevant, and more will be said about them in later chapters.

The crucial fact is that very many organic molecules can be formed and they can be quite stable, and they can have many shapes while still retaining a covalent structure. Add to this that the series of weaker bonds can be part of a larger structure and contribute to its catalytic properties. For divine providence to have preordained, if that is the right word and concept, then these laws of chemistry together with its underlying physics are enough to generate life. No matter how the laws arose and even if they are the rules of only our universe, it is enough to make life form and evolve here.

REARRANGING COVALENT BONDS

The interatomic energies holding different covalent bonds together can vary. Consider the following reaction:

A—B + C—D   A—D + C—B.

It involves an exchange of partners. The free energies of the four bonds will determine the equilibrium constant. The concentrations of the species in a reaction mixture, and the equilibrium constant determines which way and how far the reaction will go from its initial state in a given closed system. How fast it will go, however, is another question. Its speed depends on the energy of activation or the amount of additional energy, which will have to become temporarily associated with the bonds for the groups to dissociate and allow for the possibility to reform in a different combination. This additional energy is thermal energy. Such energy by chance either enters a bond as the result of bombardment of the molecule by other molecules or is reduced by bombardment of other objects by it. This chance exchange is the reason that temperature is so important for living systems. At high enough temperatures there is no function for an enzyme. A forward and back reaction will occur because a particular reaction can go in both directions, because the energy of activation is sufficiently small. Therefore, there is no cellular control. At low enough temperatures, the reaction will not advance at all or only go very slowly. The important condition for life is that some reactions may not proceed because of lack of a catalyst, but certain catalyzed reactions can occur, in the same environment, if energy coupling and appropriate catalysts are present.

CATALYZED REACTIONS

A variety of substances can catalyze reactions: metals, organic chemicals, and enzymes. How do they do it? Basically, they bind the substances, affect the charge distribution, which lowers the energy of activation, and the reactants go over the "energy hill" to liberate products. A quite common case is that in which A—B reacts with an enzyme, say EH, and forms E—B; and A is bound to a group (usually a proton) abstracted from the enzyme, yielding AH. Then the C—D molecule reacts with the enzyme to form C—B. Finally, the enzyme recovers its H group by rebinding AH, and A is liberated from the enzyme by binding D to yield A—D. The enzyme is back in its original state, here designated as EH. While the involvement of a hydrogen atom is common, it is not obligatory. All of this business is much faster than the uncatalyzed reaction. Moreover, and very importantly, it can be much more selective and occur more rapidly.

DRIVING A CHEMICAL REACTION

If a cell "needs" to have a compound formed, but the spontaneous chemistry works the other way, then seemingly the needed compound would never be formed; and the molecule, if it were formed, would tend to fall apart. How can the cell accomplish its desired goal? The answer has to be that the cell must couple the "needed" reaction to another reaction so that the second one drives the needed one. Most typically, reactions are "driven" by another reaction that, if the driving reaction spontaneously occurred, would dissipate its free energy, thus wasting the energy from our anthropomorphic view. Only if the two reactions are mechanistically connected is there effective coupling. In other words, the driving reaction can only go if the reaction to be driven actually takes place. This coupling requires that the two processes are somehow linked. While enzymes are catalysts, they are more than just that because they can be constructed not to carry out either of two reactions independently, but to occur only if both enzymatic functions occur at the same time.

Chapter 2
Preamble to Life

The conditions that were necessary for life to begin on earth are reviewed; primarily these were the prerequisite chemicals and prerequisite processes for the origin of First Cell. Once arisen, it could grow, divide, and its descendants could evolve. Although the combination of events that actually started life is only dimly understood, it is evident that a simultaneous coalition of mechanisms had to function simultaneously and in the same place for life to be "kick-started". In an abiotic world these must have included the numerous and fruitless formation of "informational" macromolecues that could not carry out needed functional processes because of lack of support, but eventually there must have been a way to utilize these informational molecules to construct others. These would actually function as catalysts and probably were ribozymes, which are molecules that are almost enzymes. These could alter available small molecules and form macromolecules and do fruitful work. The third absolute requirement is the existence of a capability of transducing energy into a utilizable form that flexibly could enable (favor) the synthesis of small molecules and macromolecules and carry out other energy-requiring cellular processes. But useable energy alone would initially lead to nothing productive. The life-generating event had to do with the simultaneous existence of these three particular processes within a single lipid vesicle. Of course, it also required the availability of sufficient and appropriate chemicals from the environment. These essential processes and organic resources could spontaneously produce molecules that would catalyze the creation of more "cells" and then bigger and more sophisticated molecules. This continuing replication and continuing evolution eventually led to diversity and the massive world biomass on the assumption that new structures and mechanisms continued to evolve.

THE ABIOTIC WORLD

In the beginning there was a "big bang" resulting in an expanding and cooling universe. As the material became more spread out in the resulting cooler milieu, aggregates of various kinds formed. The spreading was not uniform, however, because of the effects of gravity. In many parts and regions of the universe, concentrations of materials were higher. In these places high density developed as stars formed and went through their cycles. Nuclear reactions took place, and when the stars finally collapsed, some of the resultant materials were cool enough to become stable small atoms. Then the process repeated itself and the next star generation led to some still larger atoms. Finally, in the third generation of stars there was a sufficient quantity of large enough atoms so that the kind of life we currently experience on earth became possible. As a time line, our universe came into existence 13.7 billion years ago and our sun (Sol, a third generation star) appeared 4.5 billion years ago, as did everything in our solar system, including the earth. While the sun is too hot, the remainder of the solar system is cool enough to be able to coalesce into solids and molecules. The rest of the solar system includes interstellar space, planets, various comets, asteroids and planetoids. Organic molecules were made in various ways in various locations; they arose partly under the aegis of cosmic ray energy and ultraviolet light. These agents make and destroy organic molecules. However, some persisted, and some that fell to earth were probably important as resources for early life. These organic molecules were in addition to those produced on earth (see below).

Four and a half billion years ago, as the planet earth was formed, it was too hot for life. A barrage of meteorites bombarded it. If there had been any living thing to sterilize, these impacts would have killed them. There was no liquid water. It is thought that the intensity of meteorite bombardment decreased about 4.2 billion years ago. At this time surface water and oceans developed, and a variety of more stable organic chemical syntheses began to occur.

RESOURCES AVAILABLE FOR LIFE

The earth, as part of a third generation star, had adequate amounts of atoms of a reasonable size. Obviously, H, C, O, N were the key elements, but P, S are also vital. Then there are heavier elements such as Mg and especially Fe that also have a vital role. Getting all of them to form the appropriate compounds is not trivial and does require energy.

Light energy from the vacuum ultraviolet to the infrared would have been available in abundance, but other than for generating heat it would have been useless for a living creature with no ability for photosynthesis, and it would actually have been quite dangerous. When the earth cooled down sufficiently, there would have been liquid water, oceans, and rivers. There would have been organic compounds, vesicles, and reactants in the environment that had not yet reacted with each other, and there would have been atmospheric disturbances. The formation of vesicles would have depended on continual mixing by cosmic events, volcanoes, deep-sea vents, and ocean currents. The reason for this discussion is that the First Cell would have needed everything "handed to it on a platter" from the environment.

THE STAGES OF DEVELOPMENT OF LIFE

Figure 2.1 shows the entire evolutionary process; however, for this chapter it is the abiotic chemical processes shown at the bottom and in an expanded way in Figure 2.2, which presumably supplied the chemical and physical basis

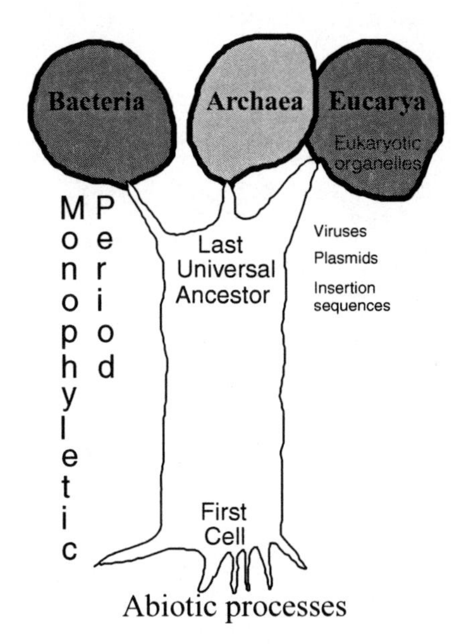

Figure 2.1 Abiotic, First Cell creation, Cellular processes, and Diversity.

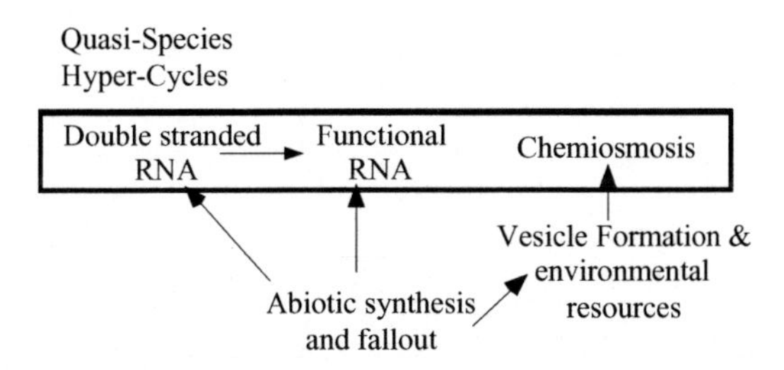

Figure 2.2 Abiotic processes in a lifeless world.

for life. For a non-photosynthetic organism, a chemical non-equilibrium situation in the environment could have been the only available source of useful energy.

THE GENERATION AND PRODUCTION OF THE ORGANIC MOLECULES NEEDED FOR LIFE

Miller and Urry (see Miller and Orgel, 1973; Chang, Mack, Miller, and Strathearn, 1983; Brack, 1998; Deamer and Bada, 1997; Deamer and Fleischaker, 1994) carried out a very important experiment. They prepared a gas mixture that

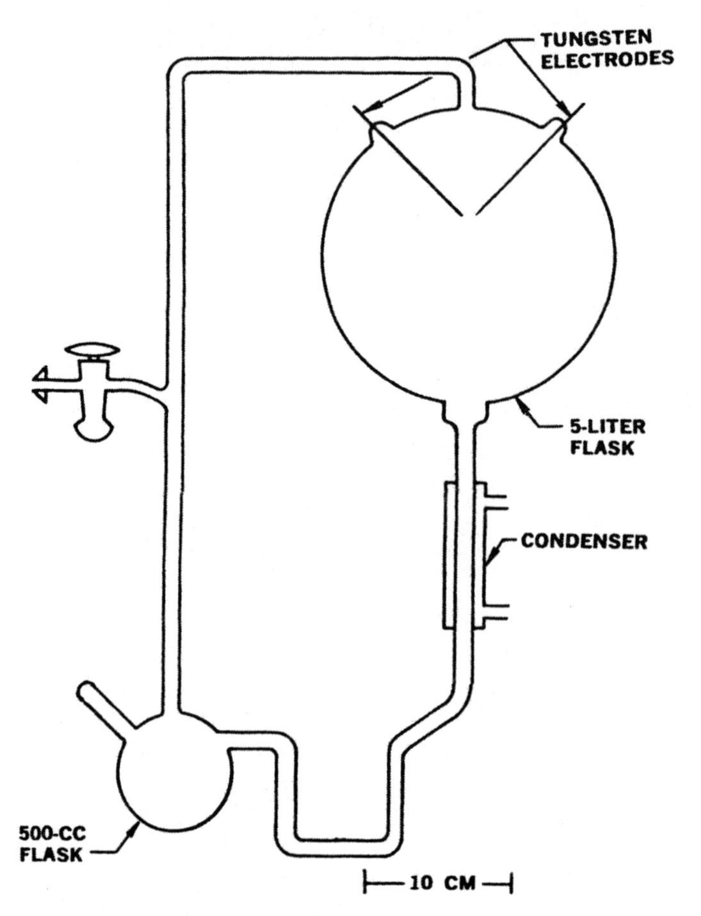

Figure 2.3 The essential aspects of the Miller and Urry experiment. Diagram taken from Miller and Orgel (1973) with permission.

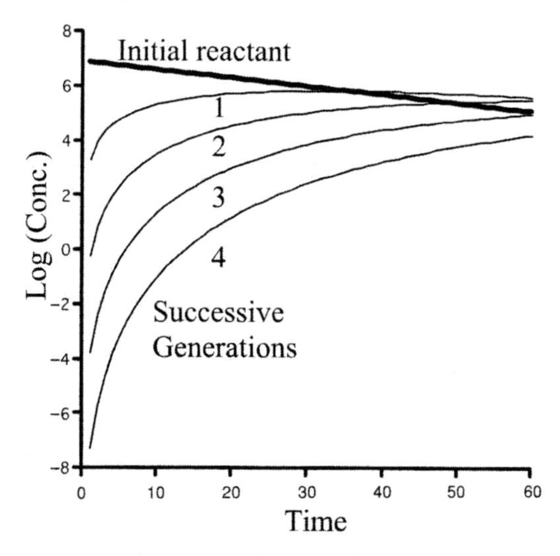

Figure 2.4 Kinetics (idealized) of successive stages of cycles of burying and unburying. Taken from Koch and Silver (2005).

they thought simulated the early atmosphere. Into this atmosphere they injected energy in the forms of electric arcs and ultraviolet rays. This energy led to combinations of very simple starting molecules and to larger ones that had more internal energy. As well however, the same sources of energy were able to break down these constructed molecules. The very clever part of their experiment was in the design of the apparatus (Figure 2.3). It was arranged so that the water-soluble compounds were condensed out of the gas phase and dissolved in the liquid aqueous phase by the condensation of the water vapor. Consequently, these molecules were no longer exposed to the destructive sources of energy. This meant that complicated molecules were made but then "buried" and not destroyed. Although the differential effect of the details of the condensation process should be important and we must assume that this part of the generic burying process is specific and selective, bigger more water-soluble molecules will accumulate over smaller ones. Molecules would be accumulated in a water phase or under ground and would have been preserved there. In the early world these accumulating resources would be around to serve for the formation of life. Figure 2.4 shows some idealized kinetics of such a process going through several burying and resuscitation stages. This would lead with time to larger and more complex structures. It is hard to imagine life arising and becoming more sophisticated without such cyclical processes. Such cycles of burial and upheaval of chemicals were needed and presumably occurred in a world experiencing many geological changes.

Chapter 3

The First Cell

The conditions necessary for life to begin on earth were reviewed in Chapter 2. The prerequisite chemicals arose abiotically on earth and in interplanetary space. We assume that abiotic processes allowed the First Cell to arise. It could then grow, divide, multiply, and evolve. The combination of abilities that actually started life is only speculation, but the mechanisms can be cataloged. In an abiotic world, they must have included the semi-conservative replication of informational macromolecules. Secondly, there must have been a way to utilize these informational molecules to construct molecules that actually functioned as catalysts (such as ribozymes, which are RNA molecules that are almost enzymes). These catalysts did alter available small molecules into molecules useful in living systems and in forming macromolecules with special shapes and chemical functions. The third requirement is a capability of transducing energy from the environment into usable forms in order to enable the synthesis in cells of l thermodynamically useful but unstable, small molecules and macromolecules, as well as to carry out other cellular processes. The life-generating possibility had to depend on the availability of sufficient abiotic chemicals and at least these three essential abiotic processes that would have, by rare chance (once, or the first time, would be enough) have produced informational molecules, which grew and catalyzed the actions that created big and sophisticated molecules able to generate what was needed for further growth.

Consequently, we can tacitly assume that the current life on this planet started with a unique event empowering at least these three processes at a particular instant of time. This event was the concatenation and propinquity of several essential processes for life (at least the three mentioned: information transfer, chemical catalysis, and energy trapping). They had to be simultaneously functional at the same time and the same place, which no doubt was a single vesicle. All these essential processes had to be working in the same vesicle for it to become a living entity. Individually each one of these several systems had, almost certainly, been created many times by chance. However, once the important "protocell" had facilities for all of the absolutely essential component processes, it became alive because it could reproduce and grow. This unique vesicle had just become the First Cell.

To repeat and emphasize, the processes in this cell must have included at least: (1) a way to transduce energy from the environment, (2) a way to copy informational molecules with only occasional errors, and (3) a way to use that information to do useful cellular work. The resulting cell was able to replicate. Even more importantly, its descendents could and did evolve. All the rest was comparatively easy—Darwinian selection of the fittest would suffice. Given adequate time, many more variants of the biological systems were developed.

ENERGY TRANSDUCTION FOR BIO-UTILIZATION

A lot of energy passes through the earth. The spectrum of energy quanta is quite diverse and the amount falling on the earth is, indeed, quite vast. It ranges from the light energy from infrared to cosmic rays and includes ionizing particles as well. Additionally, thermal energy stored in the earth is immense; its quality ranges from thermal energy at the few Kelvin degrees of the coldest place on the surface of the earth to the temperature of the iron core in the center of the earth. None of this energy serves at the "beck and call" of any living or primeval entity. All that energy is just energy, without any direction or purpose; it is certainly not "free energy".

The concept of free energy came from J. Willard Gibbs with ideas going back to Sadi Carnot, the man who first understood how a steam engine worked. The "free energy" concept originated from the existence of a purposeful machine. For example, although we may have fuel and oxygen and we can burn the fuel to produce heat, how do we trap that energy to do useful work? That is, how can we get it to do the things that "we" want done? We need and depend on many man-made mechanisms, for example, machines with pistons and cylinders and valves. In the first instance, we needed and Watt in fact built, a controllable steam engine. With this kind of machine, we could transduce the energy of combustion into the rotation of the axle and then transduce the energy formed by this rotation further to make a railroad engine roll and pull a train. With machines, man can accomplish many different things; biological systems need to accomplish many different things, too.

For the biological case, the source of the free energy must be at hand, the mechanism to transduce it must be present, and the way to store the energy in a desirable form must also be possible. In the world's biology today, respiration (the combustion of organic molecules with oxygen) may lead to conversion of ADP into the more thermodynamically unstable ATP. The ATP (and GTP) can be consumed to power protein synthesis, etc. Since proteins are thermodynamically unstable molecules, the product protein molecule would conserve some of the energy. For the first living cell, these stages must also have existed, and, most importantly, they must have existed before life arose, although the input energy was probably not in the form of ATP.

THE REPLICATION OF MACROMOLECULES

Much of the interaction among human biological organisms (e.g., political science, psychology, and sociology) obviously arose from our ability to use verbal language. This arose from its precursor of a system of grunts, etc, and from essentially symbolic languages of obvious meaning. Basically, language in all its forms is simply a code that the sender and receiver have agreed upon. These codes can have some basis in, but they need have no actual relationship to, the information transmitted. Because of its structure, the semiconservative double-stranded nucleic acid provides the facility for the copying process. A single double-strand nucleic acid is copied into two identical double strands. The actuation of this depends on the molecular forming of hydrogen bonds so that adenine uniquely pairs with thymine and guanine pairs with cytosine. It is a matter of having both hydroxyl groups and amino groups that pair and also of having one shorter pyrimidine and one longer purine that fit together to span the same length for both of the A-T and G-C pairs. Amino groups go with hydroxyl groups and long heterocylic bases go with shorter heterocyclic bases. That determines the pairing. Could it have been any other way? Of course! Uracil could have hydrogen bonded with diaminopurine, for example. Still other pairings could have served. Maybe they do serve on other planets. Clearly, however, once a system was established, no other system could work in the same environment because of the competition with the first established and at least partially perfected system. This is because the first system would have had some time to evolve and become better (more efficient). The two systems would strongly interfere with each other, and the first produced form would most usually win. Having thought about it, I am convinced that our system, with small planar heterocylic molecules like these two purines and two pyrimidines that can uniquely pair, with two or three hydrogen bonds, is the most convenient and easily developed system possible within the limitations of heterocylic organic chemistry.

THE FIRST CELL DEPENDED ON ABIOTICALLY FORMED BUT SELECTIVE CATALYSTS

Catalysts are molecules that reversibly bind to an appropriate substrate, effectively lowering the energy of activation so that the substrate molecules are converted into something else. It is possible that at higher temperatures the same reaction could have taken place without an enzyme. On the other hand, higher temperatures would have lowered the needed energy of activation for all possible reactions; whereas, the catalysts at a lower temperature could be more selective. This is a key feature of living systems: they carry out certain reactions and do not carry out others.

Consider the Universal Catalyst. This is a non-existent, fictitious, and metaphysical construct introduced to make a point. If it did exist, would its amount increase or decrease with time? It would catalyze its own synthesis and also its own destruction and every other reaction as well. The answer has actually been alluded to above. Its amount must decrease because the free energy of hydrolysis of the bonds of the molecule would favor their hydrolysis in preference to their synthesis in an aqueous environment. It is tacitly assumed that most organic catalysts are complex molecules, especially biocatalysts.

The take home lesson is that a catalyst, even a complex catalyst like the ribosomal assembly for protein synthesis, must be selective. Of course, for protein synthesis the selectivity is determined by the regulation and function of the messenger nucleic acid, and without mRNA, the modern protein synthetic system creates nothing. Platinum Black is an inorganic catalyst with some selectivity; it can function to hydrogenate many organic molecules. Many other inorganic catalysts exist and have more restricted and selective abilities, but enzymes are very special and in some cases uniquely specialized.

The crucial catalyst for the First Cell must have functioned to link some nucleic base derivatives together to form a chain. These, currently unknown, molecules had to react spontaneously because the energy to drive the reaction came from the free energy of the coupling reaction linked to the growth of the chain.

The selectivity to yield a specific nucleic acid chain had to come from a double stranded intermediate, so that a sequence of bases in a chain would be reproduced (copied) because of base pairing. There is much more to be learned about this.

THE THREE MOST ESSENTIAL PROCESSES CARRIED OUT BY THE FIRST CELL

In this chapter we have listed the minimal set of systems to constitute a living cell. They are

(I) a source of free energy in the environment that can be trapped by the cell and a mechanism to drive particular chemical reactions in the opposite way from the direction they would spontaneously take;

(II) an ability to form a macromolecular chain in which the individual units could be any of a number of constituent molecules and a way to use an existing one as a template to create a double-stranded molecule;

(III) an ability of the chain molecules to catalyze reactions that are of value to the cell.

This is strict Darwinism, but in the next section we will see some more of the details. The important consideration above was that these three systems could have been enough for life to start. Moreover, growth provides ways for life not only to perfect these three systems but also provide ways to develop many more systems present in living creatures today. Some of these aspects are considered in the next chapter.

Chapter 4
Development of Cell Physiology and Diversity

After the First Cell arose, mutations occurred and sometimes the mutant organisms would grow and prosper. Prospering would mean displacing the parental type, but it would also mean displacing other mutants and other organisms if they were not quite as fit. Selection of the better organisms is called (in addition to the "survival of the fittest") the "competitive exclusion principle of Gause". This latter phrase is used particularly when we talk of competition between species. Most students of evolution believe that almost all biological development took place under the auspices of Gause's principle. I believe that the stringent form of this rule controlled evolution until after the time when the bulk of the many cellular processes in the world today were developed that are possessed by typical cells. Before this, because of the strong competition no more than one stable species existed at a time (in a habitat). Consequently, much evolution of cellular mechanisms took place before the time that stable diversity arose. The almost unique organism present in the world at this time is variously identified with a variety of terms, such as the Last Universal Ancestor (LUA), the progenote, the cenancestor, and several others. The point is that there was before this time one species, which need not be given a name, although "prokaryote" would be suitable. At the time of the LUA there were one world species and one mutant of it that evolved into the domain of Bacteria. Around this time there also came the formation of the Archaea/Eukarya precursor, and soon the parental prokaryote population was eliminated by competition. The time of these events is about 3.0–3.4 billion years ago. This is to be compared with the time that the earth cooled enough and the number of meteorites hitting the earth slowed enough for life to be able to start. That time was at about 4.1 billion years ago.

Development of cell physiology from the time of the First Cell to the time of the LUA must have been a slow and arduous process. There were many processes and systems to be made functional and effective and this had to have taken many millions of years.

In a pre-sexual world with few resources and limited metabolic capabilities, the argument of Gause is almost certainly correct and replacement by the single most effective new form would have occurred over and over again when the habitats overlapped. But how did the world's biota ever become diverse? The fact is that permanent diversity arose only after a sophisticated and complete system of cell physiology arose. This basic cell physiology with variations is common to all organisms today. The initiation of stable diversity is not at all trivial. The diversification of a monophyletic world containing only the single organism, that is the best to have evolved up to that time, required a mechanism leading to the reduction of competition of new mutant forms versus the old. Of course, it is the best (and most successful) organism evolved so far.

Therefore this leads to the proposition that, for evolution to lead to the diverse Domain of Bacteria and its separation from the precursor of the Archaea/Eukarya that led to the Domains of Archaea and Eukarya, something very special had to occur. There had to be to a newly arisen basic problem that faced all extant organisms at that time. At that time, this one problem must have been independently solved with two different biological solutions. It was this combination of events that opened up new ways to grow without competing with other classes of organisms. The probable advance that led to the splitting off of the Domain of Bacteria, I believe, was the development of a strong enclosing wall, called the sacculus. This allowed the innovator that first formed it, and its descendants, to survive when the cells' biochemical success caused the internal osmotic pressure to become too large. However, as important as this advance was, it would not be enough to create diversity. That required another advance that occurred at near the same time to overcome the same problem. In our world this led to the development of the Archaea/Eukarya line. This favored their growth and prevented their displacement by Bacteria as the latter came to be more abundant and became able to occupy new niches and, of course, vice-versa.

CLASSES OF METABOLIC SYSTEMS THAT HAVE DEVELOPED IN THE LIVING WORLD

New metabolic processes were developed by the descendants of the First Cell to allow organisms to cope more effectively and reproduce more successfully (Koch, 1994). We come back below to the classes of systems that are characteristic of modern cell physiology. However, the enormity of the various biochemical mechanisms and of the cellular physiology that developed after the time of the First Cell and before the time of the Last Universal Ancestor must be considered here. Figure 4.1 brackets these developments between the start of life and when it started to have stable diversity.

SEMI-CONSERVATIVE REPLICATION AND TRANSLATION

Semi-conservative replication is vital for information copying, but the mechanism for translating the nucleic acid language into a protein code is very

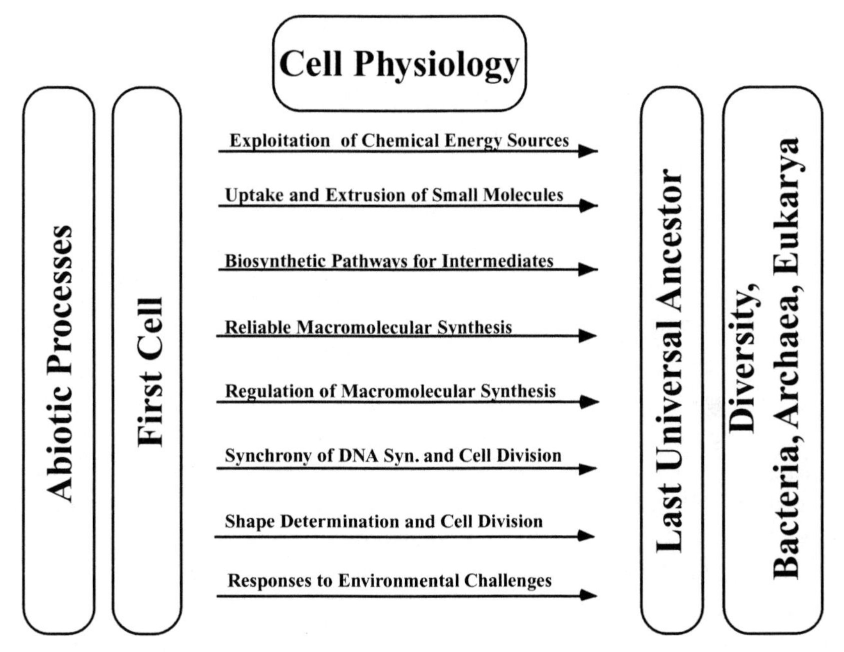

Figure 4.1 The global systems that were perfected before stable diversity developed but after the First Cell arose.

much more arbitrary. Seemingly, it could have been much different and still worked. The codification deduced by Francis Crick (which turned out to be exactly correct) is that there are 64 combinations of three successive positions in the chain of nucleic acid bases. These are called codons and determine an amino acid. Surely, chance events had to do with the establishment of the code in the first place, but clearly once it was established, there was no changing or turning back. On the other hand, the more theorists consider the "code", the more they are able to see a rhyme and reason for its detailed arrangement.

ONE MUTATION AT A TIME TO DEVELOP MANY DIFFERENT METABOLIC SYSTEMS

Life in the very early world must have been limited by the inputs of organic molecules and most critically the ability to extract usable energy from the biosphere. Selection and evolution had to be based largely on this latter single, effective, stringent, criterion. The cells would have only gradually acquired the flexibility to convert one intermediate into another, so there would be little ability for different organisms to specialize on particular resources and have little ability to use alternative energy sources. Under such circumstances the argument of Gause would necessarily apply, and replacement by the most effective life form present would occur when the habitats overlapped. Then how did the biota ever become diverse within a single region? The finding, which is not so surprising, is that permanent diversity arose only after the broad basis of cell physiology had arisen. This array of many physiological abilities is common to all organisms today. Comparing organisms, there is small to large variation in detail but not with the goals and purpose of the physiological systems. Pre-LUA evolution was, therefore, linear in that an improvement wiped out all the "not quite as successful organisms."

Because the initiation of stable diversity is not at all trivial, it follows that for the evolution of the domain of Bacteria and its separation from the evolution of the other two domains, there had to be a quite different situation: a qualitatively different problem had to occur that had probably never occurred before. A circumstance which could lead to diversity is that a new basic problem arose to face the then living organisms and that, at least, two different biological solutions to overcome this challenge developed in different organisms. Moreover, these advances needed to develop nearly simultaneously. Such multiple and different developments opened up new ways to overcome the limiting problem at hand and also allowed the multiple forms to grow non-competitively with each other.

Possibly, the advance that led to the splitting off of the Domain of Bacteria was the development of a strong enclosing cell wall, called the sacculus, to protect the organisms against osmotic pressure. This development allowed the individual innovator, which first formed it and its descendants, to survive when the cells' biochemical success caused the internal osmotic pressure to become too great.

However, as important as this single advance was, it would not be enough to create diversity. That required a separate but quite different advance only shortly before or shortly afterwards, leading to the Archaea/Eukarya line. Their advance must favor their own growth, even though they had acquired a high osmotic pressure, and also prevent their displacement by Bacteria as the latter began to be able to create and occupy new niches, and displace previous populations. An advance for the Archaea/Eukarya, I believe, was production of pseudomurein. (While some Archaea bacteria have a sacculus made of pseudomurein, others and Eukarya today have different methods to solve the same osmotic problem.) These three Domains, then, completely eliminated earlier prokaryotes which were either unable to cope with the internal high turgor pressures or could not successfully survive by maintaining a lower turgor pressure and consequently were less metabolically efficient.

EVOLUTION DURING THE INTERVAL BETWEEN THE FIRST CELL AND THE LAST UNIVERSAL ANCESTOR

A great deal had to happen between the time a self-reproducing life form successfully first arose and the later time when stable diversity became established by its descendants. In this interval, many processes had to be created by evolution and fitted into many enzymatic systems. By definition, there was no stable diversity until the time that two functional systems actually arose; this was at the time of the Last Universal Ancestor. One might have expected an early radiation of biotypes starting from the very beginning of life, but this did not happen. It is usually suggested that the evidence for absence of such an early stable radiation is that there is so much commonality in the processes presently expressed in all living cells of all domains. This is taken to mean that the hypothetical "LUA" contained and contributed the set of vital systems into all three Domains at a quite late point of evolution when there was only one Kingdom or Domain present. Thus, the LUA must have already possessed the same set of abilities, as essentially all modern cells do, although not as well functioning as in present organisms. Major evidence for the absence of stable diverse genotypes earlier is the available DNA sequence data that supports a single root for the tree commonly identified by Carl Woese and his group as the LUA. Evidence for the commonality of origin is that the sequences of a gene isolated from different organisms in different Domains or species are similar for many different genes. Of course, there are some clear, detailed differences; these are variations and alternatives within the examined genomes. These differences can be justified because there have been several billions of years up to the present for variations to develop from the time of the LUA.

For a long time before the LUA, evolution and selection were occurring on many fronts for many different genes. However, evolution was linear; first an improvement in one gene succeeded another and replaced the parental type gene and the population, and then an improvement in another gene occurred and replaced it still again. No doubt short-lived diversities were developed and then eliminated, or sometimes they succeeded and generated a new population. This kind of development is the hallmark of linear evolution. Any improvement would lead to replacement of the previous strain and possibly to elimination of an improved variety of an independent gene present in that strain, a phenomenon called periodic selection (Atwood *et al.*, 1951, Koch, 1972). This must have happened billions of times. Probably, it contributed to the long interval between the First Cell and the Last Universal Ancestor.

THE DEVELOPMENT OF CELLULAR PROCESSES DURING THE ERA OF PROGRESSIVE EVOLUTION

Woese, Kandler, and Wheelis (1990) proposed that in the time when the LUA lived, the domains of Archaea, Bacteria, and Eukarya arose. Apparently, all three developed in a geologically short period of time. The major point of this chapter is that diversity *did* wait until the major processes of cell metabolism, cell regulation, etc were functional. Collectively, this means that thousands of genes evolved to function in these systems. In each system, a number of enzymes and proteins must have operated sufficiently well and were able to work together. Moreover, systems were evolved to function for a large number of diverse purposes. Therefore, millions or billions of mutated organisms arose and competed against other organisms long before the stable diversity of the modern Domains developed. (Please note that all of this happened a long time before the beginning of the fossil record. This is important because of the fixations of the creationist and the paleontologists on bones.)

The myriads of genes functioning in a primitive cell can be categorized and organized into eight divisions (Koch, 1993, 1994) (see Figure 4.1). The first six are classes of household functions that serve in the following ways:

(i) to generate energy in a form that could be used to cause transformation of small molecules, the syntheses of macromolecules and to do osmotic work;

(ii) to accumulate (transport) needed substances from the environment into and to export waste products from the cell;

(iii) to chemically manipulate available small molecules to make the ones needed for macromolecular syntheses;

(iv) to replicate and repair informational DNA (and RNA);

(v) to regulate the synthesis of proteins and nucleic acids appropriately and synthesize macromolecules accurately via selective protein syntheses, reading of messenger RNA molecules, and degradation of RNA and proteins;

(vi) to synchronize DNA syntheses within cell division events.

While improvements in these phases were taking place, there was a seventh category that was being improved, also at many levels, and which serves an important vital role:

(vii) to regulate the amount and kinds of primary informational molecules in the individual cells.

An additional eighth general and broad category was developed which serves:

(viii) to resist sporadic environmental challenges (Koch, 1993).

Obviously, in the world today an organism has to be quite adaptable to persist. It seems unlikely that the earliest life forms had the adaptive capabilities of modern organisms; therefore, the selective pressure to develop such versatile protective mechanisms must have been great.

Each category from this list of the major headings may consist of many processes, and each should contain a number of gene products. They are all important, and some functions of each must occur for life to become as extended and expanded as it is today. Thus, in a third general kind of argument, we can suggest that a very large amount of genetic development had to take place in many aspects of biology before stable diversity ensued. This would take a long time.

CONTROLLING THE REDUNDANCY OF GENETIC ELEMENTS

While all above categories are crucial, the seventh category of controlling the redundancy of genes in a cell is critical for the evolutionary process itself. At the most fundamental level, if a cell arises by division and fails to have an essential genetic element, it is doomed; on the other hand, if there are too many copies there will be a decrease in growth rate due to overproduction of the product it specifies and thus create an imbalance in the cell's economy. This may be lethal; at least, it would decrease the competitiveness of the organism (Koch, 1984a, 1985).

There is an additional reason for not having an excess of copies of a necessary functional gene: multiple copies will drastically slow evolution in the Darwinian "mutation-and-selection" sense. With multiple copies of a gene present within an organism, a single advantageous mutation increases the fitness of the accompanying unimproved versions of the same gene simultaneously. The cell containing both a few copies of the new version and a number of copies of the older version may be less strongly selected when competing with cells containing only the non-improved multiple copies than if it had only improved copies. Therefore, although the cell having a single copy of the improved version, in addition to other unimproved versions of the same gene, may eventually outgrow the original cell type, it will grow more slowly than if the non-improved gene copies did not tag along.

Consequently, there would have been strong selective pressure to evolve life forms with a general strategy of having only a single gene copy of a structural gene (Koch, 1984a, 1985). Of course, the working copy had to be capable of efficient function so that just a single copy could accommodate the maximal needs of the organism. This is important because it can be shown (Koch, 1979) that unless a gene, when fully turned on, can serve the maximum needs of a haploid organism in its current environment, then tandem duplicate gene copies will, sooner or later, be generated and maintained in the population. Of course, duplication, refinement, and specialization will take place to create new functions and be the basis of evolution of major changes (as a usual paradigm in evolution, but also see Koch, 1972).

Although much has been written about "selfish" DNA, it must be noted that selfish DNA could only prosper later in evolution. This is both for metabolic and genetic reasons. Early on a selfish gene could only replicate within its current host and be incapable of immigrating to other sites within the genome of its current hosts or emigrating into other hosts. It also needed to do something favorable for its host. For such processes to be possible, the host organism had to be sufficiently sophisticated to carry out versatile genetic processes and to insure that the extra genes and gene products would not have decreased its fitness significantly, since other factors would have become more critical to its biology. At some earlier time in the evolutionary process such genes would have lowered the fitness of the host's genome and this would have been also self-destructive of the "selfish" DNA.

The ratio of gene activities of various kinds would have been critical. Therefore, selection for mechanisms precisely regulating the replication and transcription of genes in order to coordinate their numbers with others had to occur. This must also have been a very powerful evolutionary force. A great growth advantage must have been provided to some very early life form when it finally acquired the fourfold combination of (1) forming one covalently linked chromosome containing all the kinds of needed genes; (2) controlling the initiation of the replication of that chromosome to a favorable stage in the cell cycle; (3) regulating chromosome replication to coordinate it with the cell's success in biomass formation; and (4) regulating the function of many genes at both the transcriptional and at translational levels. Once such a single linkage group and regulatory systems were achieved, growth would have been very much more efficient. Subsequently, a loss of any one of these features would have been very disastrous in the presence of relatives that had not lost them. In particular, if division occurred more frequently than replication, then lifeless cells would be produced, thus decreasing the rate of production of viable cells to below what it optimally could take place. In the opposite case, if there were extra copies of the chromosome linkage group, then an altered cell physiology would have been appropriate; and either very large free-living cells requiring major intracellular transport processes [or long cells of a narrow cylindrical shape] or cells with a higher DNA content than is optimal in relation to other cell constituents would result. These possibilities would waste resources, slow growth, and also interfere with the evolutionary process because of the existence of multiple gene copies.

MEETING ENVIRONMENTAL CHALLENGES

A very important problem for living organisms to face soon after life started is compensating for variations in its surroundings. These variations would include temperature changes and the affects of photons and ionizing radiations. These factors would vary from time to time and place to place. Additionally, nutrients come and go, and hydration is variable. As a result of these factors, a cell needs to develop special techniques, methods, and strategies. All of these belong in Category viii of the above list, and all are vital.

BIOENERGY BEFORE THE LAST UNIVERSAL ANCESTOR

Why did stable diversity not develop earlier? One possibility could be that the major cause of the lack of the development of long-lived diversity before the LUA, in part, resulted from the dearth of energy resources.

What energy sources for biological purposes are used today? The major categories will be listed in the order that they have been suggested in the earlier literature for the first source used by living organisms. However, I will divide these possibilities according to the organismal type:

- Bacteria
  - Fermentation
  - Respiration
  - Photosynthesis
    - RC1 á là the heliobacteria
    - RC2 á là the green bacteria
    - O_2 photosynthesis á là cyanobacteria
      - With both RC1 and RC2
    - RC1 á là the proteobacteria
- Archaea
  - Methanogenesis
  - Respiration
- Eukarya
  - Respiration
  - Oxygenic photosynthesis

None of these are reasonable as the first source of energy; they are all too complicated, require accurate protein synthesis and sophisticated synthesis of small molecules, like chlorophyll and seem to have evolved after the LUA.

What other sources might be tapped? That is, what else is there? What has been suggested?

The utilization of the poly-phosphates of rocks;
Some exotic form of photosynthesis, no longer existing;
A type of chemiosmosis not seen in today's biota.

Note that I have not included the rhodopsin-based photosynthesis of halobacteria.

These all may appear very unlikely. There is no question that rock polyphosphate could yield energy from its reaction with water. Even though the rocks are still here, there is no life form today that gains its "daily-bread" from this type of geological biochemistry. If there has been a quite different kind of simple photosynthesis, it no longer exists. The process used by halobacteria is simpler than the chlorophyll based mechanisms listed above, but it is still complex and requires the synthesis of rhodopsin and retinal. An even simpler mechanism would need a photoreceptive molecule to adsorb the light without destroying itself, transform the energy into the separation of charges, somehow prevent the recombination of the plus and minus charges, and have a way to utilize the charge separation, for example, to make a protonmotive force.

There is more to be said in the favor of chemiosmosis. The process happens in many ways in many cells. The basic idea of Mitchell, for which he garnered a Nobel Prize, is that if charge is moved across a lipid bilayer, work must be done to accomplish this. Then that work is stored up across this membrane as the developed charge separation. Such an assembly can be thought of as a charged electrical condenser. This stored work can be recovered in some other form from the membrane by letting the charge separation become discharged under controlled conditions that causes some other reaction to occur. If properly done, this can afford energy, even free-energy, that can be used by the cell for various purposes. Probably, the best-known example is the metabolic generation of a voltage across a bacterial cell's cytoplasmic membrane or across a mitochondrion's inner membrane due to respiration and the coupling of the reverse flow to the conversion of ADP and P_i to ATP. But that is not the only example. Although such chemiosmosis fires up much of the world's cellular metabolism, there are many other types. There are chemautotrophs that oxidize various reductants with oxygen and in so doing create a protonmotive force. This energy is used for cell processes, for example, for the reduction of CO or CO_2 to the oxidation level of formaldehyde or formic acid, needed for synthesis of cell substances. In a similar manner, there are organisms that use H_2 to reduce oxidized substances. This energy powers up much of cellular biochemistry. There are two types: one uses O_2 as the typical oxidant together with various reductants, the other uses H_2 with various oxidants. In the world today, both are actually dependent on the oxygen-producing photosynthesis carried out by other organisms to produce both the oxidant and reductant. Photosynthesis on earth is the big energy coupler of environmental solar energy into the biosphere. However, that is not the point in this chapter and this discussion was an aside to, "What was the first source of bioenergy?"

If, for example, an energy source were generally limiting, then variations in temperature, water content, nutrients, etc. would be less important factors. As a consequence, selection for a better energy production mechanism would be paramount. As mentioned above, the other possible early energy sources might be from phosphate anhydrides or a currently unknown kind of photosynthesis, but neither could have been very extensively used. I do not believe that fermentation or a primitive photosynthesis of either the Photosystem type II or type I could have contributed at all before the time of the LUA. The photosynthetic mechanism, because of the physics of trapping light energy, would have required sophisticated light absorbing molecules and associated proteins which would have required extensive evolution. Moreover, it appears that photosynthesis of existing kinds arose only after the creation of the Domain of Bacteria and after considerable evolution of diverse organisms belonging to that Domain had occurred. Additionally, the fermentation mechanism is unlikely to be an early process because many enzymes would be needed and fermentable organic molecules would have been too rare in the environment.

Any of these possibilities (a novel photosynthesis, chemiosmosis based on chemicals available in an anaerobic world, and abiotic pyrophosphate utilization) if they had existed would be of limited use to creating world biomass, because they could function only to the extent that the needed reactants were simultaneously available to a cell, able to react, yet not having reacted previously. The probable reaction couple suggested by Wächtershäuser (1988a, 1998b, 1990, 1993, 1994) is the formation of iron pyrites from ferrous ion and hydrogen sulfide and evolution of H_2 (or equivalently the simultaneous reduction of some organic molecule). Above it was discussed how this could have been the energy source in a chemiosmotic type reaction that could have started life. Its role during the long time needed for the development of cell processes (the monophyletic period) is also likely. The reductant half reaction is

$$Fe^{++} + 2H_2S \rightarrow FeS_2 + 4H^{++} + 2e^-$$

The oxidant half reaction could consist in the consumption of these two excess electrons in the reduction of carbon monoxide or dioxide to form formaldehyde or formic acid. While such a total reaction would occur with a large yield of free energy, no chemotroph known today uses this system. In today's world few niches exist where all three reactants are abundant. For the more typical chemiosmostic mechanisms used by chemotrophic bacteria today, the yield of free energy is less, there are only two instead of three reactants, and the role of the oxidant and reductant is clearer and simpler. Also, the niches for the known autochemotrophs may be more common when an adequate amount of a relevant oxidant and reductant are simultaneously present and chemically able to exergonically (spontaneously) react. Of course, if the oxidant and reductant are in a common environment, then the requirement is that they have not already reacted and been used up. This means that the oxidant and reductant have been newly mixed together, otherwise they may have already done so. In modern times, habitats where this is so generally exist because of availability of the products in today's vast biosphere of oxidants and reductants, which are almost exclusively the spin off from the oxygenic photosynthesis process.

The conclusion to be drawn is that the iron pyrites type of chemiosmosis mechanism may have functioned before photosynthesis but supported only a very small world biomass, and now is non-existent. In a world after the LUA, the oxidizing and reducing reactants are continuously generated by photosynthesis metabolism, which would favor simple redox reactions.

The other potential energy source suggested to contribute to early life is the anhydride bonds in phosphate-containing rocks. Because the energy transfer carrier most important in living systems (adenosine triphosphate) is crucial, we should not totally exclude consideration of this abiotic analogue of ATP. Phosphate anhydrides of a variety of kinds would have been present on the surface layers of the earth. They would have been generated thermally, and the energy of the anhydride bond would be, in principle, available to be used in a coupled reaction. It could happen if polyphosphate were split via a cellular catalyst in a way to drive an endergonic (energy requiring) reaction in the non-spontaneous direction and in a way useful to the cell. The barriers to doing so are immense in several ways. First, there must be propinquity of the needed three or four reactants. These chemical species are: the phosphate anhydride reactant; the organic substrates to be altered; the catalyst, and the cell. The inorganic phosphate species is highly charged and outside the cell while the substrate and catalyst are usually inside the cell. Additionally, in a wet environment, hydrolysis of the anhydride may have occurred spontaneously, dissipating the source of energy. Also, if alternative reactants were present (inside or outside the cell) the pyrophosphate bond might be split and its energy dissipated without producing the component needed by the cell or by making transferable bioenergy. Finally, the ability of an organism (or its contained catalyst) to favor the reaction adequately above the abiotic, spontaneous rate could be limiting by an energy source.

Many phosphorylation reactions happen as a large part of current intermediary metabolism in living systems when a phosphate anhydride, such as the pyrophosphate groupings of the ATP or one of its analogues, is split and one residue transferred to the enzyme. Then in a second step, the phosphate is transferred from the enzyme to the molecule that becomes activated by adding this group. This, in turn, can later react, liberating the phosphate compound and carrying out a needed cellular process. This strategy is the primitive process, but later on in evolution, with more sophisticated enzymes, only one reaction instead of two occurs to do the job. Enzymes called "synthases" or "synthetases" do this combined process. These enzymes bind ATP and split the anhydride bond; and simultaneously, but at another site on the enzyme, remove the elements of water from the two molecules, say a hydrogen from an amine and an OH from a carboxyl-containing group of the other, to form a peptide or amide bond between them. By looking over metabolic reaction pathways, the conclusion can be drawn that synthases are relatively late evolving events in intermediary metabolism. From these considerations, it probably is a reasonable conclusion that the use of phosphates was a later development than the chemiosmotic process.

To repeat, methanogenesis and or single-stage photosynthesis could have supplied much larger quantities of energy for life than the more energetic but primitive iron pyrites chemiosmosis mechanism would have been able to produce, but both of the former processes are very sophisticated and, therefore, in principle could not have arisen early. Thus, the early world was energy limited, and quite possibly the source of available free energy was this special iron pyrites form of chemiosmosis.

The big change in the earth's biosphere was the one increasing the biomass and hence increased the opportunity for diversity; it depended on the development of new energy sources—photosynthesis and methanogenesis. One must emphasize again that both are very complex processes and would have required extensive evolution. Besides serving as energy sources, they would also provide utilizable carbon resources. The other forms of trapping light energy (anoxygenic photosynthesis of several kinds), however, would not. It is not clear when each evolved; probably methanogenesis evolved in Archaea after, or shortly after, the time of the LUA. This also would be true of the halobacterial type photosynthesis, while anaerobic or anoxygenic mechanisms for photosynthesis developed in Bacteria, probably considerably after the time of the LUA.

Throughout most or all of this period, biological growth depended on the input of nutrients from abiotic, terrestrial sources and from extraterrestrial fall out. However, only near the end of the period before the LUA could it have occurred by recycling materials from dead or living organisms. Saprophytic growth could only happen when secretion of enzymes from cells became possible. Consequently, an alternative to perennial energy limitation would be a consequence of perennial limitation of carbon at intermediate oxidation states between CH_4 and CO_2. In all of these cases the competitive exclusion principle should have functioned to yield an almost pure monoculture.

THE GAUSIAN PRINCIPLE AND ITS RELEVANCE

Further aspects of Gause's (1935) principle are appropriate here. The relevance of the lack of supply of various energy sources or the shortage of a supply of diverse carbon compounds containing oxygen and/or nitrogen follows from the modern form of the theorem of Gause. Gause's "Law of competitive exclusion" states that the maximum number of species in a habitat depends on the number of available homologous resources that different species can specialize in and consume. It is a variant analogous to Darwin's survival of the fittest but applies to different species instead of variants within a species. The idea is the same in both cases. If one individual of a line, strain, variant, or species grows better (faster) than others, even if the advantage is slight, given enough time the more rapidly growing will displace the less rapidly growing. The displacement will be complete unless there are habitats where the advantages are negated or reversed. The modern formulation of the Gausian principle is that an ecosystem can only support as many species as there are numbers of independent resources or sub-habitats. Then various extents of specialization will allow co-existence.

From the discussion presented earlier about the pre-LUA world, a possibility might have happened, but actually did not. This possibility is that there could have been diversification, for example, by one line using the chemiosmotic energy source and the other using a presumed poly-phosphate energy source. Similarly, if nutrients were the most limiting for growth, then certain lines might have specialized in glycine consumption and others on alanine, if these two were the major available metabolizable sources. According to the 16s rRNA phylogenetic tree, none of the possible stable diversities developed earlier than the LUA, and only then did the diverse evolutionary branches became stable. Consequently, I take it that a single factor, most likely the chemiosmotic energy source, was the dominant need and the other possibilities were virtually irrelevant. On this basis, evolution proceeded with many improvements in various characteristics but mainly to economize on energy and become more efficient, even if only very slightly reducing consumption of the single main limitation.

LINEAR EVOLUTION VERSUS BRANCHING EVOLUTION

Under the auspices of the extreme version of Gause's theorem, only a single variant should exist in a constant environment with a single limitation, although a variant could be replaced by another and then another and another. Under only slightly different conditions and circumstances, a version of this rule would lead to a series of coexisting species (see below). Gupta's analysis (Gupta, 2001, 2002, Gupta and Griffiths, 2002) of prokaryotes is different from the 16s rRNA tree from Woese's laboratory in that Gupta's pictures the linear development of different major groups of bacteria from the First Bacterium. Gupta's model of evolution postulates that low GC Gram-positive bacteria arose from a currently non-existent class of cells, called above the pre-LUA or prokaryote cells. After the emergence of Bacteria, there were two types of development. One was the development of diversity within these Gram-positive bacteria. This diversity was stable because of the variety of niches and resources, which had become available because the biosystem had become more diverse and not limited in the same restrictive way. The other was the macromutation of a single individual in this group (or actually in a subgroup of the low GC Gram-positive cells). This individual was the founder of a new group. These two processes, diversification within and founding a new group without, happened again and again. From the Gause principle it can be surmised that the new group arising from a radical advance, if it occurred in a completely uniform environment with a single limitation, would eventually displace completely the precursor population. What actually must have happened is that there was enough of the old habitat and conditions that the old clades (group of related individuals) persisted, and the advanced form was able to evolve and adapt and then re-radiate but did not completely replace the old. Then eventually a new "macromutation" led to the next advance. For example, all the Gram-negative species come late in the Gupta sequence, suggesting that only after plants and animals developed was there an important niche for bacterial species that could adapt to inhabiting these living new "worlds" as a symbiont, commensal, or pathogen.

MODEL FOR THE DEVELOPMENT OF DIVERSITY

Consider two sister variants; imagine they both grow at different rates and start with different initial numbers. The numbers of the original type, N, and a mutant, M, will change according to:

$$N = N_0 e^{\mu_n t} \qquad \text{and} \qquad M = M_0 e^{\mu_m t},$$

where the subscript zero designates the initial numbers; and μ_m and μ_n symbolize the specific growth rates of the two varieties. Since e, the basis of natural logarithms, was invented 500 years ago, it has been known that the ratio of M/N will grow larger or smaller indefinitely with time, t, depending on difference of the growth rates. Only if the two rates are exactly identical will the ratio stay the same. What mathematicians have known for hundreds of years can easily be demonstrated with a computer. While this ratio changes from generation to generation it never becomes zero or infinity, however, at some point we can "decide" that one species has been eliminated.

With a chosen criterion, say $M/N = 10^{20}$ for exclusion of N, then given M_0/N_0 and $\mu_m - \mu_n$, we can compute how long it would take to replace N by M by this criterion. From a geological perspective, even if $\mu_m - \mu_n$. is very small, the time to exclusion would not be very long. If the M cells had another genomic property, a characteristic marker different from one in N which had nothing to do with the property that made M grow faster than N, then almost all M cells in the replacement population will have this characteristic marker property. Now, if a new mutation occurs in an individual cell that happens to be an improvement, then a successor species or variety will replace the current population but still have the M mutation. This type of event must have happened repeatedly, billions of times during the period of development of cellular physiology between the First Cell and the Last Universal Ancestor. This is an application of the logic behind the Gausian exclusion principle, and the above treatment is a good beginning. We need to add two more aspects, however.

The first aspect is that further diversity generating mutations would take place in the M population, dividing it into M_1 through M_j. If one of these, say M_x, is the one that contains an innovative "macromutation" of significance, then the successor population that arises by eliminating M_1 through M_j with the exception of M_x, will be characterized by the incidental changes that made all the different M's different from the N cells, but in addition the M_x population will contain some new unique changes. Consequently, the founder M_x cell that gives rise to a new population will incidentally have the changed characteristics that the N population had different from its predecessors. It will also have those changes associated with the origin of the M, and also with the incidental properties of the M_x that gave rise to a new group of organisms. The second addition to the basic "exclusion principle" is that there may be special locations or conditions where the older version can survive. This kind of persistence has been mentioned in passing above and is that cells of the old kind may persist and not be fully excluded in the greater population as the biological world became more diverse. Under these condition varieties, either if selected against or not selected for, will actually persist and may not be eliminated.

Part 2

Wall Structure

Chapter 5

Covalent Bonds and Tensile Strength of Materials

A very important physical principle that is pertinent to biology and to geology, but is also pertinent to the working of many man-made structures and machines, is the concept of "the strength of materials". When a rope, for example, is under enough stress it will rupture and it will do so in a direction at right angles to the stress and, importantly, at the weakest point. This is the precursor to the principle of Griffith, a key concept from the central branch of mechanics. However, Griffith's principle applies to solids and it extends to, and includes, the cell walls of bacteria. From the stresses involved it is clear that in order to function properly the sacculus must be held together totally by covalent bonds (and not by weaker bonds) and that the structural arrangement must be such that not too much stress is placed on any particular bond. The idea that a chain breaks at its weakest link is the older idea and applies to a one-dimensional structure. However, for a two- or three-dimensional structure, a breakage event leads to a redistribution of the stresses that increases the stress to a nearby bond and thus favors its rupture. The eventual result is a crack propagating through the solid, destroying its integrity and unity. The bacterial growth strategy must avoid such destruction. This chapter expands on this idea.

STRENGTH FROM BONDING AND STRUCTURE

The bacterial wall is much stronger per unit thickness than the walls of higher plants. This extreme strength is due to its chemical structure and to the network of connections of the unit parts that are covalently bound together to form a strong fabric. Contrast the murein fabric of the wall with a phospholipid bilayer. In the latter, the fatty acid chains group together in leaflets because the molecules are attracted to each other by weak apolar and van der Waal's bonds. Also, for the same reason the two bilayers come together and the two leaflets of the bilayer form closed vesicular structures. Apolar forces are simply the resulting repulsion between hydrophobic and hydrophilic molecules. This repulsion actually results from attraction of the hydrophilic molecules to each other because they can lower their energy by forming hydrogen bonds to hydrophobic molecules that have few or no electronegative atoms. Electronegative atoms, like oxygen or nitrogen atoms are typically present in hydrophilic molecules. Consequently, the non-polar parts are forced away from the polar parts. Van der Waal's forces also cause weak attractions; these arise due to the interaction of electrons in the negative charge cloud around H and C atoms. Such interaction results in the clouds becoming mutually distorted in a way to effectively attract the two hydrophobic molecules. Both apolar and van der Waal's forces are relatively non-directional and much weaker than covalent bonds.

A covalent bond is formed when two atoms together bind two electrons, one of which "belongs" to one nucleus and the other electron "belongs" to the other nucleus. Each atom in the intact covalent bond would have an orbital with these two electrons oriented with opposite spins. Such a bond in a vacuum, if broken, would yield two free radicals. Not so for the weaker apolar and van der Waal's bonds.

Such paired electrons are needed to satisfy the Pauli principle of physics. A consequence is that a covalent bond is much more stable than the two free radicals formed by splitting it. Such free radicals arising within a molecule in an aqueous environment would cause a water molecule to split in such a way that the OH would bind to one and the H to the other radical, eliminating the free radicals and forming only covalent bonds. It therefore follows that splitting a molecule's bond in an aqueous environment to create two molecules is easier than in the absence of water where forming free radicals would be obligatory.

One more bond type needs to be considered: the hydrogen bond. A hydrogen atom is the smallest atom and for this reason has a property not possessed by larger ones. A hydrogen atom only forms this type of hydrogen bond when attached to an electronegative atom, such as oxygen or nitrogen, and when close to another electronegative atom. For this type of bond the electron cloud from both electronegative atoms interacts with the hydrogen atom. This causes the electronegative atoms on both molecules to be attracted to each other because both are attracted to the same intervening hydrogen atom. Of course, the hydrogen atom "belongs" to the electronegative atom that possessed it in the first place, but it is a hydrogen bond because both atoms are attracted to the same hydrogen atom.

VISCOELASTIC PROPERTIES AND TEARING OF SOLIDS

In terms of its chemistry the two-/three-dimensional bacterial wall structure is essentially quite similar to the structure of vulcanized rubber, lucite, polystryrene, PVP, or Bakelite. In the bacterial wall, as in these materials, the atoms link together to form an extremely large macromolecule where all the atoms are interconnected by a network of covalent bonds. The integrity due to the covalent bond structure leads to a mechanical strength that maintains cell shape. This type of structure is much stronger than the cellulose or chitin present in plant or insect or crustacean walls. In these cases the molecules are formed of one-dimensional chains wound around each other and the chains hold together by hydrogen bonds. These materials should be thought of as cables, which do give great strength under tension, however in the side-to-side direction they are held to the other chains only by the weaker hydrogen bonds and the interwoven nature of the chains. When tension is applied only along the main axis of the fiber, they can locally and temporarily become dissociated or ruptured (particularly in an aqueous environment) and then the chains may slide and creep quite readily, one against the other, to reduce the stress and the hydrogen bonds reform.

Consider tearing a piece of paper. The paper may resist a certain amount of tension without giving much, but then forms a single tear when the tension becomes too great. If the paper is dampened, it stretches first and then tears more readily. Of course the cellulose chains are just as strong either wet or dry, but paper is formed of cellulose fibrils oriented in many directions and matted together. The fibrils in the paper are held together by hydrogen bonds. When a hydrogen bond breaks and tension is present, chains move relative to each other and finally a tear starts. When wet, this process occurs more readily because, as stress-bearing hydrogen bonds are broken, new non-stress-bearing bonds form briefly with water molecules so that the mechanical elements can move relative to each other more readily, permitting creeping movements with negligible energy cost. Hydrogen bonds can then reform, linking a saccharide or peptide of one chain with that of another, but the linkage now is in a different place than originally so that the strands have slid relative to each other. This latter process is part of the creeping and viscoelastic extension that then reaches a limit after the creeping process has allowed many cellulose chains to reach the end of many others, and then a more permanent tear commences.

In the example of the piece of paper, only one tear develops. This fact, i.e., that stress favors a single fracture of materials, is central to Material Sciences and is a restatement of the Griffith (1920) "propagation of cracks" theory. The theory proposed that when a material is subject to an even force, the stresses would be more intense in some regions than in others because of imperfections. Bonds in the vicinity of regions where there are flaws in the initial solid are those that will rupture first. When they fail, rupture occurs and the stresses become redistributed. Now some bonds will feel less stress while other nearby bonds are subject to intense stress and then they, in turn, will quickly fail. This leads to the formation of cracks, which relieve the stress in the bulk of material, stops other cracks that were incipient, and causes the fragmentation of the solid by "brittle fracture".

From measurements of cellulosic materials (say a piece of wood) containing oriented fibrils stretched until it breaks, it is possible to demonstrate that wood samples, even when dry, fail much more readily than expected from the properties of diamonds which have the same carbon-carbon bonds, but at a higher degree of perfection. (Mark, 1943). In principle, a cellulose chain should be as strong as a chain of C-C bonds in a diamond. In fact, there is propagation of cracks starting at weak points that are part of the biological structure that developed during growth of the wood. These overstressed points then fail and, like a series of dominos, the whole structure fails. Electron microscopy of wood suggests that covalent bonds seldom fail, but the regions between fibrils and cells do. More often in fact, even the weaker hydrogen bonds holding chains together often remain intact. It is the imperfections due to the detailed anatomy of the wood that determine just where the ruptures will take place.

MATERIALS WITH WET STRENGTH

The bacterial wall material, like the diamond, has "wet strength" because of its reliance on covalent bonds. Of course, there are also hydrogen bonds linking chains, but only a few compared with starch or cellulose or DNA. This is, in part, because of the paucity of hydroxyl functions on these saccharide chains. Hydrogen bonds can form between the peptide structure and the saccharide chains, as well as other molecules composed of peptide chains or other wall components. However, many of these would be broken and reformed with water as a partner when a wet wall is stressed mechanically. As our understanding increases, these hydrogen bonds no doubt will be important in quantifying the elasticity of the wall. The major point is that the cross-linked nature of the bacterial wall explains how the rupturing effects of cell turgor pressure are resisted in objects with such thin walls, but the conundrum is understanding of how the bacterium can grow and its walls enlarge without rupturing.

WALL SYNTHESIS AND THE MAKE-BEFORE-BREAK STRATEGY

Wall enlargement involves two processes: the cross-linking of nascent oligopeptidoglycans and the cleavage of old wall. This concept has been appreciated since Weidel and Pelzer's seminal paper of 1964. For the safety of the cell, the two processes of synthesis and cleavage must be well coordinated. The formation of the cross links is an interesting example of biochemistry and will occupy most of the next chapter, but the cleavage is no less interesting and very important from the point of view of controlling bacterial growth. Although the wall must be cleaved to be able to insert new material for enlargement, if weakened it is in danger of ripping or tearing. This could lead to the destruction of the organism. Although Weidel and Pelzer assumed that cleavage occurred first, we (Koch, Higgins, and Doyle, 1981) have taken the reverse position. We feel it is axiomatic that in order to keep the wall strong during growth, new wall must be incorporated and cross-linked before existing stress-bearing wall is cleaved. We call this ordering of the events the "make-before-break" stratagem. Different bacteria, however, carry out this stratagem in different ways as will be delineated in later chapters. Safe enlargement requires the strategy to be essentially foolproof. For the mechanism best understood, i.e., the strategy of growth of the sidewall of the Gram-positive rod-shaped bacterium, it is very simple and very clever indeed. The mechanism can be fooled, however, and the cell can destroy itself. Although the enzymes that do the cleavage have other important roles, their destructive role is the reason that gave them the name, autolysins.

INTRODUCTION TO THERMODYNAMICS

Covalent bonds, such as C-C, C-O, C-N, are strong and resist stretching and breaking compared with apolar and H-bonds. However, with enough stress they can be ruptured. For our context, "ruptured" usually means a forced hydrolysis of the biologically significant bond with a molecule of water. This allows the two biological parts to separate so that a chain is now split into two fragments. Of course, one end now has an added H and the other, an added OH. In the absence of stress the bond will not fail (or spontaneously hydrolyze in an aqueous environment) because the atomic forces holding the atoms together are sufficient unless the temperature is too high.

For the same reason, two amino acids will not react to form a peptide and a water molecule; i.e., there is energy of activation to be overcome no matter whether the bonds that form are of higher or lower energy than the original ones. Therefore in biological systems in order for the synthesis of a peptide or glycoside bond, a source of energy is needed. An energy-rich compound supplies the energy of activation, which besides contributing energy for bond formation is partially dissipated as heat. Note that mechanical stress can provide the energy of activation needed for hydrolysis and thereby speed the splitting of a stressed macromolecule.

There is some difference in the bond energy of the various bonds and this too can be an important factor. For the disaccharide bond and the peptide bond versus their hydrolytic products, the products are somewhat more stable (about 3 kCal/mole) than the reactants. In the reverse direction such as for example, protein synthesis, four high-energy pyrophosphate bonds supply the energy of activation for formation of a single peptide bond and the energy incorporated in the bond. In doing so, much of their energy is largely dissipated as heat. Of course, the higher the energy so dissipated, the higher the speed of the kinetic process can be.

Whether saccular enlargement or dissolution is under consideration, the problem is the reverse but not too different. However, it is more complex than for bonds between two simple molecules where only one dimension is relevant. Now the stresses are in both dimensions of the saccular surface because of the internal turgor pressure. Consequently, new insertion of disaccharide penta-muropeptide units must overcome or somehow avoid the stresses. For dissolution, the stress lowers the energy of activation needed for rupture and aids the autolysins acting on stress-bearing saccular walls.

CHEMICAL BONDS

Chemical bonds identify the linkage between two atoms. A bond can be indicated with a line on paper or rods connecting balls that signify atoms, but of course, they represent the interaction of the attractive force between electrons belonging to two different atoms and the net positive charge on the other atom. Covalent bonds are quite stable; those such as C-C, C-O, C-N, are strong and resist stretching and breaking, compared with apolar and H-bonds. However, with enough stress they can be ruptured. For our context, "ruptured" usually means a forced hydrolysis of the biologically significant bond in an aqueous milieu with a molecule of water. This allows the two parts to separate so that one molecule has become two and one H_20 has been consumed. A chain of sugars, amino acids or nucleotides, for example, is now split into two fragments. Of course, the elements of water have been added and one end now has an added hydroxyl group and the other, an added proton. In the absence of stress the covalent bond will not fail (not rupture into atoms) or spontaneously hydrolyze (in an aqueous environment) because the atomic forces holding the atoms together in a covalent bond are sufficient unless the temperature is quite high. That is to say that the energy of activation for hydrolysis is higher than the energy in the biochemical bond in question at almost all times. Certain chemical arrangements have a much lower energy of reaction than others. Thus bonds of peptides, of glycosides, or of sugar-phosphates have a much lower energy of activation than a C-C bond and this is the reason that much molecular biology is involved with polymers made of units that end (generally on both ends) with these hydrophilic types of bonds.

Two amino acids will not react to form a peptide and a water molecule spontaneously for lack of the energy of activation. This, for the same logical reason, is true for the reverse reaction as well when hydrolysis should take place. It must be that random collisions of molecules must be sufficiently energetic to cause the bond to have temporally enough energy to be equal to, or higher than, the energy of activation for the molecular bond to interact to where they may produce the new molecular arrangement. This is no matter whether the bonds that are to be formed are of higher or lower energy in totality than the original ones that are broken. Therefore in biological systems, in order for the synthesis of a peptide or glycoside bond to occur, a source and a way to introduce energy is needed. Most often an "energy-rich" compound supplies the energy of activation. This energy, besides contributing energy for bond formation, is partially dissipated as heat. Note that mechanical stress can provide the energy of activation needed for hydrolysis and thereby speed the splitting of a stressed macromolecule.

There is some difference in the bond energy of the various bonds and this too can be an important factor. For the disaccharide bond and the peptide bond versus their hydrolytic products, the products are somewhat more stable (about 3 kCal/mole) than the reactants. In the reverse direction, such as for example protein synthesis, four high-energy pyrophosphate bonds (from ATP and GTP) supply the energy of activation for formation of a single peptide bond. In doing so, much of their energy is largely dissipated as heat. Of course, the higher the energy input, the faster the dissipation of the kinetic process.

THE THERMODYNAMICS OF BOND FORMATION

A reaction event if left to itself would occur in the direction that was spontaneous and would result in a more stable bond. Such reactions, of course, require no exogenous energy source. Thus proteolytic enzymes split peptide bonds with the introduction of a molecule of water. The role of the enzyme is to effectively lower the energy of activation, but does not transfer or absorb energy to or from the reactants. The hydrolysis products of a protein are more stable than the original protein. This stability permits the process to occur without energy from the outside. To go in the synthetic direction is quite different. This usually requires coupling to an energy source that can transfer its energy to the needed task and must include the bond energy of the bond that will be formed. Most typically in biological systems this energy comes from the hydrolysis of ATP. Of course, the hydrolysis of ATP can occur spontaneously. In an energy coupling situation, the cleavage of ATP leads to the formation of an enzyme linked to a part of the ATP and in a second reaction the energy is transferred as this bond is cleaved and whose energy is shared by the chemical reaction, building up a biochemical that the cell needs. Not all energy transduction occurs exactly in this way, but all follow the rule that the energy source is dissipated only if synthetic energy needs are of a lower energy than that donated by the driving reaction that forces the synthesis. This transfer requires a mechanistic coupling afforded by the detailed arrangement of the biochemical enzymatic system.

For intermediary metabolism, conditions are usually such that reactions generally go to near completion. Thus, for biomolecules the free energy of the process is a couple of kilocalories and the equations of thermodynamics indicate that only one molecule in a million or so will remain unhydrolyzed at equilibrium. Consequently we usually do not need the full Gibbs equation of chemical energetics that calculates for a given mixture which way the reaction will go and how far it will go. But because we will need it at some point, it will be stated here:

$$\Delta G = \Delta G^{o} + RT\ln K - RT\ln Q$$

This can be read that the actual free energy, ΔG, is equal to the standard free energy plus RT times the natural logarithm of the equilibrium constant for the reaction, K, minus a similar looking expression where the actual concentrations of products and reactants in the system under consideration are inserted into an expression, now called Q, that looks like that for the equilibrium constant, K. K is the products of the products divided by the products of the reactants. The standard free energy, ΔG^{o}, is the energy involved if a mol of reactant becomes a mol of product when the concentration of all species is at the agreed upon standard values by a committee of biochemists. These are typically 1 molar, although certain components may have special values (e.g., the pH may be that of the system being used).

The Gibbs relationship concerns overall reaction and not the individual bonds made or broken. So we can apply it to the formation of a functional wall. The energetics involves the process of disaccharide penta-muropeptide synthesis in the cytoplasm, movement through the cytoplasmic membrane, linkage into a glycan chain, crossbridging of the muropeptide to another muropeptide, and finally the stretching of the nona-muropeptide. In this calculation, we must take into account the number of ATPs used in the cytoplasmic biochemistry, the UTPs used in making the disaccharide, but not the bactoprenol acting catalytically and cyclically in the export process, but including the hydrolysis of the D-Alanine from the end of the penta-muropeptide.

While I cannot carry out this calculation for murein precisely now, I can well imagine that this energetics situation may lead to a specific class of antibacterial compounds that are not under consideration today. Although I must add that the nona-muropeptide stress model introduces another energetic factor because a penta-muropeptide can only interact with an established-linked nona-muropeptide if the latter has been stretched adequately (see below).

Chapter 6
Structure of the Fabric that Covers a Bacterium

A cell surrounded only with a phospholipid vesicle (the cytoplasmic membrane) should imbibe water and swell because of its mechanical weakness. The cytoplasm of the first descendants of the First Cell probably did not have a high enough turgor pressure to cause such problems, but by the time that much physiology had been developed and the Domain of Bacteria was about to be initiated, cellular systems must have improved, tending to increase turgor pressure and nearly bursting the cell. During earlier evolution, as organisms improved their intermediary metabolism and their membrane transport capabilities, the turgor pressure would have increased and the tendency of the cell to rupture, or "blow-up", in a dilute aqueous environment would have progressively increased. Consequently a system had to be developed that was able to prevent such swelling and must have increased the cell's resistance to swelling due to its osmotic pressure. This was accomplished by the evolution of a sacculus surrounding the cell.

How could the cells protect themselves against swelling up due to osmotic stress and bursting? The obvious solution would be to create and add a strong fabric, totally enveloping the cell, which would protect the three-dimensional cell that is otherwise enclosed only by a weak cytoplasmic membrane surface. The murein of Bacteria and the pseudomurein of Archaea are the only monomolecular fabrics that have been developed in biological systems, which have sufficient strength (i.e., that have adequate numbers of covalent bonds) to protect the cell effectively against a small osmotic pressure with only a thin (single monomolecular layer) of stress-bearing wall. Plants and animals use alternative strategies for the same purpose. However, the murein of bacteria is a two-dimensional polymer formed of covalent bonds that are sufficiently strong. In it the glycan chains in a local region go in one direction and the peptide chains go nearly at right angles to form a strong surface. To form such an external covering, there are a number of quite difficult biochemical problems that had to be solved and consequently the biological construction of the external sacculus is not at all trivial. However, the difficulties were overcome several billon years ago at the time of the LUA, and Bacteria and some Archaea have evolved stress-bearing walls from their primordial wall-less precursor cell. The creative trick to achieve this biology was to copy a number of genes that were then possessed by the primitive cell, modify the duplicate version, and combine these into a functional set to create a dedicated system whose purpose is the construction of a safe, strong, intact wall. These are, no doubt, quite different purposes than those that the original genes had before duplication. Evolutionary developments in five original quite separate and diverse areas must have occurred in pre-LUA cells before the Domain of Bacteria arose.

In this case again as in the creation of the first cell, a successful combination of dedicated processes was needed. To create both the First Cell and the First Bacterium an original combination of processes was essential. This form of evolution is quite important and has even been given a special name—exaptation. The major purpose of this chapter is to describe the bacterial wall, its function and its manufacture.

THE PENTA-MUROPEPTIDE DISACCHARIDE

The basic "building block", or repeating unit, of most bacteria is a penta-muropeptide linked to β-D-N-Acetyl Glucosamine-(1-4)-β-D-N-Acetyl Muramic (or NAG-NAM). This penta-muropeptide is the typical form used in both *E. coli* and *B. subtilis*. In proper chemical nomenclature, it is:

L-alanyl-D-isoglutamyl-*meso*-diaminopimelyl-D-alanyl-D-alanine.

Diaminopimelic acid is abbreviated both as DAP and mA_2pm. Each muramic acid residue contains as an integral part a D-lactyl group that forms an amide bond with the amino end of the penta-muropeptide. The key point about these disaccharide penta-muropeptides units is a result of their chemistry. They are constructed so that they have four relevant terminal functional groups: two functional hydroxyl groups (the 1 and 4 hydroxyls of the two ends of the two sugars of the disaccharide), an amino group (part of the zwitter group) on the end of the diaminopimelic acid of the penta-muropeptide, and a potential carboxyl group in the middle of the D-alanyl-D-alanine group, which is exposed once the terminal D-alanine is removed by transpeptidation. Then two penta-muropeptides can form the nona-muropeptide tetra-saccharide cross-bridge. There are three places for attachment of two other units. This is the minimum number needed to form a fabric or wall covering from any structural unit. Moreover there is a fourth attachment site. The role of the fourth is not as a permanent attachment site but, even though only temporary, it is vital for the growth process (see below).

THE CROSS-LINKED FABRIC-LIKE STRUCTURE OF THE BACTERIAL WALL

The molecular structure for the five-membered peptidoglycan unit of *Escherichia coli* and *Bacillus subtilis* was deduced thirty years ago with much critical chemistry (see Morin and Gorman, 1982; Ghuysen and Hakenbeck (ed.), 1994; Koch, 2000, 2001). The structure was modeled with the aid of a computer program (Koch, 2000, 2001). This disaccharide penta-muropeptide structure is shown Figure 6.1. Except for the wall-less mycoplasma, it is typical of Bacteria generally. There are chemical variations in different species, but functionally they are all the same. For all bacteria, many identical units are linked together to form the sacculus that surrounds completely the growing bacterium.

The carbohydrate chain portion of the unit shown is a disaccharide formed via a β(1-4) glycosidic bond of N-Acetyl-Glucosamine (NAG) with N-Acetyl-Muramic acid (NAM). The D-lactyl group within the NAM is linked to L-Ala-D-Glu-*meso*-A_2pm-D-Ala-D-Ala. (I have used a different set of abbreviations than given above, but the formula is the same and as given in Figure 6.1.) This is quite different from peptides in usual proteins. It contains D-amino acids;

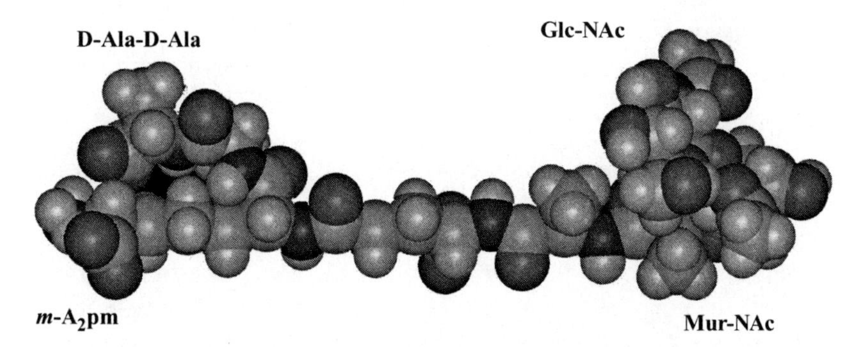

Figure 6.1 The penta-muropeptide of *E. coli* and *B. subtilis*. The partially stretched molecular molecule (Koch, 2000) is shown. The right-hand side ends in the disaccharide portion. These sugars allow it to be inserted and to extend the glycan chain. The left-hand side has two ends. One branch ends in a D-Ala-D-Ala-. By transpeptidation it can enter into an amide linkage that is almost identical to an ordinary peptide linkage. The other branch ends in a zwitter ion, which is composed of a carboxyl and an amino group. The amino group from one penta-muropeptide may interact with the penultimate D-Alanine from another penta-muropeptide to form a tail-to-tail bond to cross-link the murein. The central region consists of amino acids in peptide linkage. Some of these amino acids are special because they are D instead of L. One amino acid is glutamic acid linked at its γ position instead of the usual α linkage.

it links glutamic acid through its γ-carboxyl group; and contains an exotic diaminopimelic acid residue:

$$H_2N\text{-}CH(COOH)CH_2CH_2C(COOH)NH_2.$$

The important point about this molecule is that its conformation is that of a zwitter ion. In a zwitter ion a carboxyl group ionizes its proton and the neighboring amino group acquires the proton to yield a net zero charge, even though it is a doubly charged molecule.

Most of the amino acids within the muropeptide are connected via their α-amino group to the α-carboxyl group of the adjacent amino acid in the way found in normal proteins. This is true of one of the ends of the diaminopimelic acid in murein linkage, but not the other. Consequently its chemical structure in the penta-muropeptide is:

$$RHN\text{-}CH(COOR)CH_2CH_2C(COO^-)NH_3^+$$

Here the R's represent the rest of the amino acids and the right- hand side has the zwitter ion. It is the -NH_3^+ that is important for the murein structure in forming the tail-to-tail cross-bridge. It is this tail-to-tail bond that allows the murein to cover a two-dimensional surface.

The fundamental chemical advantage of having both carbohydrate and protein components in the basic unit is that it allows the development of multiple covalent cross-links that can form two- and three-dimensional structures. Both ends of the disaccharide are used to extend the carbohydrate chains in the wall polymer. On the peptide part, there are a number of functional groups. For joining the two peptides together in the case under consideration there is one carboxyl group from the D-Alanyl-D-Alanine and there is one amino group from the *meso*-diaminopimelic acid. They are linked together in the cross-linking process. While this was said above, it is necessary to note that there is another free amino and another carboxyl group remaining after two penta-muropeptides link to form the nona-muropeptide (and liberate a D-alanine). These free groups serve an important role in the growth of the cell wall (see later). The polymerized structure should be thought of both as a stress-resistant fabric and as a fisherman's net. But the emphasis must be on the complete enclosure of the cell and its ability to resist stretching beyond a certain degree.

OTHER MUREIN WALLS

The vital chemical feature of the cross-linking process is that the penta-muropeptide is able to form a cross-bridge with another such peptide by a tail-to-tail arrangement. Making the tail-to-tail linkage depends on the special structure of the penta-muropeptide. Usually this is accomplished by the presence of a special polyfunctional amino acid with one more amino group than is present in most naturally occurring amino acids (Tipper and Wright, 1979; Schliefer and Kandler, 1972). Typically, the diamino acid is *meso*-diaminopimelic acid as described above, although in other bacterial species it can be a different but quite functionally similar molecule. It is most generally a diamino acid. This serves the role of providing an extra functional amino group to cross-link the carboxyl group of one peptide to the extra unusual amino group of the second unit. In some bacterial species, the third amino acid in the chain, instead of being diaminopimelic acid as shown in Figure 6.1, can be L-glutamate, L-homoserine, or L-alanine. In these three cases, the tail-to-tail linkage is made but depends on the extra α-carboxyl group in the D-glutamic acid in the other chain.

ENERGY FOR WALL SYNTHESIS

The amino group of one penta-muropeptide chain is attached by the action of a membrane-bound transpeptidase to the carboxyl group of the penultimate D-alanine of a nearby D-Ala-D-ALA peptide in exchange for the terminal D-alanine. The inclusion of an extra terminal D-alanine during synthesis is the feature that allows the peptide bond to form outside of the cytoplasmic membrane. During ordinary protein synthesis inside the cell, peptide links are formed using the energy of the high-energy phosphate bonds of ATP and GTP together with the information from an mRNA. It is done on ribosomes and requires many additional factors. Since the energy supply from the triphosphates is not available outside the cell proper, it is necessary to provide the energy built into the disaccharide penta-muropeptide as a disposable peptide bond before it is secreted through the cytoplasmic membrane. It is this disposable bond that participates in the transpeptidation step that is critical to the formation of the bacterium's exoskeleton.

This formation of tail-to-tail bond is the Achilles' heel of the bacterial wall growth mechanism and is the site of action of many useful antibiotics (which is why all these topics are included together in this book whose ultimate goal is analyzing the role of wall antibiotics in inhibiting bacteria).

To summarize the special chemical features of the bacterial wall: In the peptide portion, there are some D-amino acids instead of the L-amino acids commonly found in proteins. D-glutamic acid occurs in an uncommon linkage, i.e., through its γ-carboxyl instead of its α-carboxyl group. In many, there are species of diamino acids that do not occur naturally in proteins. Besides these chemical features serving necessary roles in wall formation, coverage, and function, an additional advantage of such a structure is to make the peptidoglycan resistant to many of the proteolytic enzymes widespread in nature.

WALL SYNTHESIS AND THE MAKE-BEFORE-BREAK STRATEGY

Wall enlargement involves two processes: the cross-linking of nascent oligopeptidoglycans and the cleavage of old wall. For the safety of the cell these two processes must be well coordinated. Although the wall must be cleaved to be able to insert new material for enlargement (at least in Gram-negative cells), when weakened it is in danger of ripping or tearing, which could lead to the destruction of the organism. Although Weidel and Pelzer (1964) assumed that cleavage occurred first, we (Koch, Higgins, and Doyle, 1981) feel that it is axiomatic that in order to keep the wall strong during growth, new wall must be incorporated and cross-linked before existing stress-bearing wall is cleaved. We call this process the "make-before-break" strategy.

BIOSYNTHESIS OF PEPTIDOGLYCAN

In some sense, the wall growth is more difficult than chromosomal replication, RNA manufacture, or protein synthesis because the integrity of a stressed structure must be maintained during each step. Wall enlargement has many of the general properties of these other processes, but there is a feature that stands out that is quite different ; it is that the final assembly must take place at a site remote from where normal cellular regulation can be directly expressed. Another special property is that the installation of peptide cross-links occurs not by a templating procedure but more closely akin to the way that a nascent polypeptide chain spontaneously folds itself into a native, functional form with only information from its own primary sequence and the environment. This auto-control feature is necessary for wall formation, as it is for protein folding, because there is no alternative source of information. This brings us back to the major point that the last steps are biosynthetic processes that must take place outside the cell in a region in which the cell can maintain, at best, only very limited control.

VARIATION OF THE BASIC UNIT AND THE LINKAGE OF THE CROSS-BRIDGE

The basic "building block", or repeating unit of bacterial murein, is a penta-muropeptide linked to –β-D-NAG (1-4)-β-D-NAM-(1-4)–. The bond between the saccharides is formed by ordinary glycosylation, but the peptide bonds are only possible because the muramic acid residue contains a D-lactyl group with a free carboxyl group. Variations in the composition of the peptide chain occur in other organisms and can be taxonomically important. These variants may become medically important in the time to come. So it is worthwhile to outline a bit of what is in the older literature. The most extensive discussion of these themes and variations is to be found in Schleifer and Kandler (1972), Tipper and Wright (1979), and Rogers *et al.* (1980). In the scheme of Schleifer and Kandler (1972), the peptidoglycan shown in Figure 6.1 is classified as A1γ. The A indicates that the third residue is iso-*meso*-diaminopimelyl (diaminopimelic acid, abbreviated DAP or more recentlyA_2PM). Its ω-amino group is used to form the cross-bridge. The 1 indicates that there is no intervening peptide sequence, and the symbol γ indicates that the third amino acid is *meso* and not the LL form of the diaminopimelic acid. The Schleifer and Kandler scheme does not cover all cases because further substitution of the chain can occur, in particular, the amidation of the α-carboxyl group of the D-glutamic acid. In the B type peptidoglycans, the polyfunctional third residue is not used for cross-linking; instead, the second residue (glutamic acid) is used. In these cases there is always a poly-functional bridging diamino acid. This means that invariably the cross-bridge peptide is formed in the same chemical way, i.e., a terminal D-ala-D-ala bond on the donor peptide engages in transpeptidation to combine the carboxyl group of the penultimate D-alanine to form a new peptide bond with an amino group present at the end of the bridge amino acid. It can be presumed in all these cases that the bridge amino acid(s) was added before the externalization of the unit.

There are some common structural features among the various bacterial peptidoglycans. One is that the glycan chain is fairly non-polar by virtue of extensive derivatization of the hexose. The only free hydroxyl groups remaining are those on the 6-position of both sugars and the one at the 3-position of the NAG moiety. In some bacterial species there is a reduced level of acetylation on the saccharides, the chains become considerably more polar, but this deacylation presumably takes place after transport through the wall. For the structure shown in Figure 6.1, the peptide chain possesses three negative charges and one positive charge at neutral pH, but in some species, for example, *Enterococcus hirae*, formerly called *Streptococcus faecium*, there are no remaining charges.

FORMATION OF THE DISACCHARIDE UNIT

There are several stages to the formation of a murein unit: first, the synthesis of the sugars and muropeptide, then the extrusion to the outside of the cytoplasmic membrane, followed by its incorporation into the growing saccular

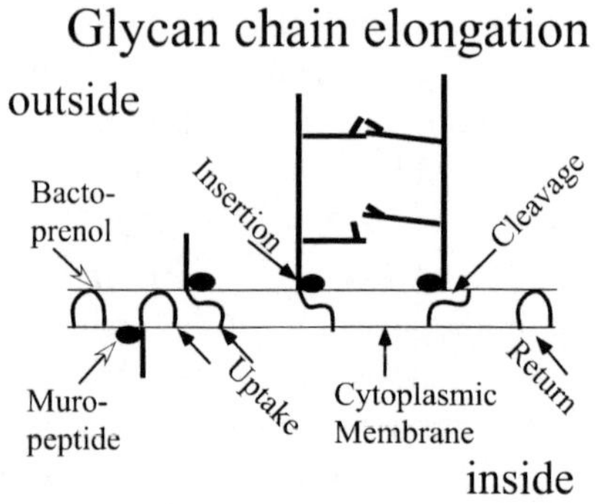

Figure 6.2 Transport, insertion, cleavage, and cross-linking. Shown from the left-hand side is a bactoprenol molecule, then a bactoprenol attached to a disaccharide penta-muropeptide facing the cytoplasm. Further to the right, after transport a bactoprenol attached to a penta-muropeptide disaccharide facing the environment is shown. Still further to the right a pair of growing glycan chains cross-linked at two places is shown. The disaccharide is shown inserted at the base of the chain and the penta-muropeptide is shown unrolled in the environment outside the cell or in the periplasmic space, both of which are hydrophilic. Sites of cleavages for the transglycosylase-mediated insertion step are indicated. The liberated bactoprenol inverts itself through the cytoplasmic membrane to carry another disaccharide penta-muropeptide through the membrane.

wall. Finally we must consider the autolysis. Except for the first and last, these are diagrammed in Figures 6.2 and 6.3

For the biosynthesis, starting from glucose, a number of steps are needed to get to the beginning of the unique part of the pathway. We may consider that the starting material for the unique branch of peptidoglycan synthesis is N-acetyl glucosamine-1-phosphate, a modified glucose that already has in place two needed functional groupings and is primed with the needed free energy in the form of the phosphate group. The hydrolysis of another phosphate from UTP supplies the energy to yield UDP-NAG. Note that the employed UTP (uridine triphosphate) reagent is the one usually reserved for carbohydrate chemistry in biochemical systems.

To this point in the biosynthesis, the biochemistry is standard, simply variations of the steps found in many metabolic pathways, however, the rest of the pathway is quite different. In the next step, UDP-NAG is involved which is needed for itself but also needed to make UDP-NAM for the cross-linking process. So a part of the NAG acquires the functional D-lactyl grouping needed for later cross-linking. The reactant in this case is phosphoenolpyruvate, a very high-energy phosphate compound. This step is made irreversible by the reduction of the double bond with NADPH + H^+ to yield NAD^+. This stabilizes the product, fixes the ether linkage, and fixes the stereochemistry as D. Now we have

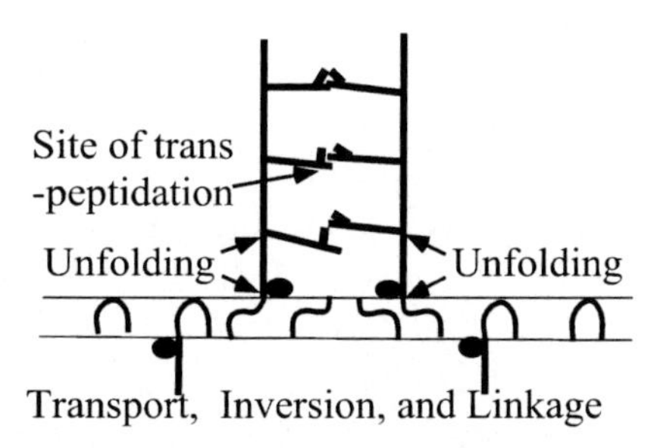

Figure 6.3 Cross-linking. Shown on the bottom, as in Figure 6.2, are the steps of transport, inversion and glycan insertion of the disaccharide penta-muropeptide. The transport of the disaccharide penta-muropeptide unit is shown moved through the phospholipid membrane and inserted into the growing glycan chain. A membrane-bound enzyme, which is a transglycosylase, inserts the growing chains. This results in the new unit's insertion into the nascent wall. The two rolled-up penta-muropeptides unroll and are electostatically attracted to each other. For the peptide cross-linking, a diaminopimelic acid end of one and the D-ala-D-ala of the another peptide chain interact and become attached via the agency of a membrane-bound transpeptidase and with the release of D-alanine. The remaining two ends interact electostatically to be part of the control system for wall growth.

a carbohydrate with an attached carboxyl group, which has a similar reactivity to the chemical reactivity of the carboxyl groups present in amino acids.

FORMATION OF THE MUROPEPTIDE PORTION

The peptide chain is formed and elongated with a series of amino acids. This, however, is not done in the way that proteins are made upon ribosomes, but rather it is done in a way closer to that used in the synthesis of peptide antibiotics and some small hormones. The normal mechanism for protein synthesis would not suffice because it could not cope with D-configurations or function in the linkage of the glutamic acid by its γ-carboxyl group. In most instances, L-alanine is added to the lactyl group; in a few species this first amino acid is glycine or serine. The coupling in all cases is done by stereospecific enzymes that are called synthases, so named because they couple the free energy of the terminal phosphate bond in ATP to force the covalent bond formation (with the formation of ADP and inorganic phosphate). Then in the same way, the amino group from D-glutamic acid is incorporated. The D-glutamic acid is made from L-glutamic acid by L-glutamic acid racemase. The next peptide bond formed is (via addition) to the γ-carboxyl group of the D-glutamic acid. This also is unusual biochemistry. Then, the amino group comes from *meso*-diaminopimelic acid, for the penta-muropeptide that is shown in Figure 6.1. This unusual amino acid is an intermediate in the biosynthesis of lysine. Since one end of a *meso* compound has the D configuration and the other L, it is not surprising that it is the L end that reacts to couple with the D glutamic acid. Then finally, D-alanyl-D-alanine (prefabricated by alanine racemase and coupled with the D-ala-D-ala ligase with the concomitant hydrolysis of ATP) is coupled to the carboxyl end.

To recapitulate: (1) individual enzymes are required for the amino acid additions for the muropeptides and the process is quite different from the protein synthesizing apparatus. Each enzyme is specific for its amino acid substrate, the configuration, and the nature of the linkage. The energy cost is paid with phosphate bond energy of ATP. (2) The peptide structure is special. The alternation of D and L configurations probably serves the functional role of creating a compact form to pass through the membrane, yet allowing an open extended configuration when it comes to be cross-linked and bear stress in an aqueous environment. (3) The universal aspect in all bacterial species is that the chain when it is completed has five amino acids. It terminates, with no exceptions, in a D-ala-D-ala grouping and the contained peptide bond of D-ala-D-ala supplies the energy for cross-link formation.

TRANSPORT AND LINKING

The process is outlined in Figures 6.2 and 6.3. In this section, we deal with the movement of the penta-muropeptide to the outside of the cytoplasmic membrane and its incorporation (insertion) into the murein wall.

After the basic unit has been prefabricated, it is delivered to the construction site for the wall. The problem is that this site is on the other side of the cytoplasmic membrane. Lipid-soluble compounds can pass through the bilayer, while polar molecules of the size of the disaccharide penta-muropeptide (molecular weight ~1000) cannot. A similar problem arises in the synthesis of periplasmic proteins, outer membrane proteins, and proteins for export. The usual solution for transport of these latter proteins is now well known. The coding regions for these proteins lead to a product with an initial hydrophobic stretch of amino acids, the leader sequence, which can pass through the membrane because it is so hydrophobic. Once threaded through, the rest of the chain with its assortment of hydrophilic and charged residues can be pulled through because, as one hydrophilic group enters, another hydrophilic group is likely to be leaving. This to some degree balances the energy requirements. Subsequently, an enzyme, called the signal peptidase, cleaves the leader sequence off.

For transport of the wall unit, the strategy is different than protein transport, but in some ways it is similar: the muramic acid muropentapeptide is attached to the non-polar membrane-bound bactoprenol carrier. Then the other saccharide, the UDP derivative of N-acetyl-glucosamine, is added to the NAM derivative to form the completed basic unit in a state already linked to the carrier molecule. At this point in some species, extra bridges such as a pentapeptide of glycine residues may be added. Charged-tRNA molecules successively donate these. Because charged tRNA molecules could be formed only on the cytoplasmic side, it is clear that at this point the basic unit is still on the cytoplasmic side of the membrane. Then a lipid-soluble carrier (undecaprenol, also called bactoprenol) embedded in membrane aids the transport. Finally the unit is dissociated from the carrier and attached, i.e., inserted between another carrier and the growing oligopeptidoglycan chain composed of a number of peptidoglycan units. After cross-linking to the oligopeptidoglycan chain the unit is cleaved from its hydrophobic handle (see below).

This dedicated hydrophobic molecule, which functions as an aid to the penetration of the disaccharide penta-muropeptide, is a C55 compound containing two trans-double bonds and has 9 cis-double bonds. It is one of a family of isoprenoid lipids that occur in living cells throughout the three domains. From the structure of cytoplasmic membrane with two layers of fatty acid chains apposed tail-to-tail, it can be estimated that, generally, it takes a span of about 36 carbon atoms to reach across a bilayer. (Think of the size of the bilayer of opposed phospholipids, each with fatty acids of 16 to 18 carbon atoms attached to the glycerol moieties. So these C55 molecules could go across and curl back through the phospholipid membrane.) Other than the double bonds on this molecule, there is only one functional group, a single terminal hydroxyl group. This hydroxyl group is esterified to inorganic phosphate, which imparts a negative charge to the molecule. In the transport cycle it is bound via a pyrophosphate bond (with three negative charges at neutral pH) to the glycan unit. After the still mysterious transport across the membrane, the unit is almost as mysteriously inserted into the growing peptidoglycan chain, not at its tip, but at its base.

Biochemically, such basal insertion is similar to the way that other extracellular polymers, such as dextrans and levans are made. An enzyme that has two binding sites forms both of these molecules; one site holds the growing chain and the other accepts the new unit. A transfer occurs that moves the growing chain to the end of the new unit. Then the binding site must alter its chemical function, or more likely, there must be a translocation step equivalent to the translocation steps on the ribosomes in protein synthesis. The details have not been elucidated yet, so we must speculate on another possibility. This is that the relevant chain-elongating enzyme must react with the pyrophosphate-linked, growing chain to split the carrier-pyrophosphate, and it is a glycosyl-enzyme that then donates the chain to a new isoprenoid-linked unit.

Much of the biochemistry of wall production is done by the enzymatic activity of penicillin-binding proteins (PBPs). These are membrane-bound enzymes that bind β-lactams and supply a number of functions needed for wall synthesis. Since some of the penicillin-binding proteins have transglycosylase activity as well as transpeptidase activity, they are probably involved at this stage. If the scenario just described applies, the undecaprenol at this point is bound to a singly esterified pyrophosphate, which in an aqueous neutral environment would have three negative charges. An unresolved question is on which side of the membrane the terminal phosphate group is removed. The conventional wisdom is that it is on the inside in order to conserve phosphate. There is no evidence on this point, but I would suspect that hydrolysis takes place on the outside so that a less polar derivative of undecaprenol with only one negatively charged phosphate has to be transported back across the membrane to the inside. The freed phosphate can then be taken up by energy-consuming systems that (certainly) the bacteria have on hand to maintain their normal supply of phosphate. To start the whole cycle anew, a fresh NAM with both a muro-pentapeptide and the UDP group attached reacts to form pyrophosphate and liberate UMP and simultaneously attach the NAM-polypeptide to the carrier.

Chapter 7

The Covalently Linked Sacculus: the Nona-Muropeptide Model

A bacterial cell wears a stress-bearing suit of armor: the sacculus or exoskeleton. In E. coli it is composed of hundreds of thousands of disaccharide penta-muropeptides polymerized together. There are even more in the Gram-positive B. subtilis. Of course, the penta-muropeptide units are held together by strong covalent bonds, as shown in Figure 6.1, and these individual units are bound together with additional covalent bonds to make the sacculus strong.

The bacterial wall can be a two- or three-dimensional covering. There are two styles: Gram-positive and Gram-negative. They both have different strategies to export a small unit and polymerize it into the growing sacculus while maintaining saccular strength. The rationale for these architectural differences comes simply because the basic disaccharide penta-muropeptide units must pass through the cytoplasmic membrane surface in both cases and then must be linked together to cover a unit of surface area in a make-before-break fashion but in different ways (see Chapters 11 and 12). Because the enzymes that create the linkages are also membrane bound, fundamentally only a two-dimensional structure can be created. Three-dimensional walls, as for example in B. subtilis, are composed as an onion is constructed by the successive addition of layers. However, there are probably some bonds linking together adjacent layers.

The ability to form a layer from the basic disaccharide penta-muropeptide units with the facilities available from the membrane-bound enzymes is quite limited. The hexoses of one unit can be tied to the hexoses of another by β(1–4) bonds to form the glycan strand. One penta-muropeptide extending from a NAM residue can be tied to another penta-muropeptide by a tail-to-tail peptide (really an amide) bond to make a nona-muropeptide that has linked together two glycan chains. This chapter presents how the organic conformational chemistry of the murein controls formation of new murein in a safe manner.

CONFORMATION OF THE GLYCAN STRAND

A constraint is imposed by the conformation of the residues that make up a glycan chain. Consider first the glycan portion by itself composed of repeats of disaccharides of NAG linked to NAM. The single bonds that make up the glucose nucleus lead normally to the boat conformation, and hexoses typically are combined through the 1 and the 4 positions on the opposite ends of the two linked hexose rings. The carbon-carbon and carbon-oxygen bonds have strongly preferred angles; and consequently, the hydroxyl groups at the 1 and 4 of the two hexoses of the disaccharide positions protrude at definite directions. Because the glycan portion of the muropeptide could constrain the form of the fabric, I modeled the conformation of the glycan chain without the muropeptide attached to it (Koch, 2000). This was done as the first part of my job in simulating the structure of murein.

Simulating the conformation of organic molecules has been greatly improved with the current technology. In fact, in the last few years it has been become so easy with available commercial programs that writing software is no longer necessary. The modeling presented here was performed using the Insight2 program (Silicon Graphics, CA). Insight2 allows the user to choose atoms and residues to construct an organic molecule, a way to invert optical activities, and a way to link groups together. It can constrain the position of atoms. Fixing the position of an atom was essential to estimate the effect of stresses in the wall causing conformational changes in the murein in the bacterial wall envelope. To begin with, the conformation was varied by a part of the program in order to search for the minimum energy conformation.

Beginning with an oligoglycan composed of eight disaccharides of N-acetyl glucosamine and N-acetyl muramic acid in β(14) linkage repeats, the computer searched for the conformation yielding the minimum energy for this pure carbohydrate structure. The computer simulations showed that the linear polymer of the 16 saccharides, when suspended freely in dilute solution, can have a variety of conformations, which are all essentially of the same energy. Of course, when a new chain is first formed by transglycosylase action and no cross bridges are yet present, none of the internal energy is due to stress. For a small elongation by stretching (~10 Å), only a little additional energy is involved, but the total energy increases rapidly as the chain is stretched to be longer and longer. There are two points of interest here: one is the range of conformations of strain-free structures with only minimal thermal energy and the other is the conformation of the oligoglycan under stressed conditions. If minimization is initiated from an initially linear structure, the conformation is approximately helical (see Figure 7.1). For clarity, the ether oxygen atoms in the

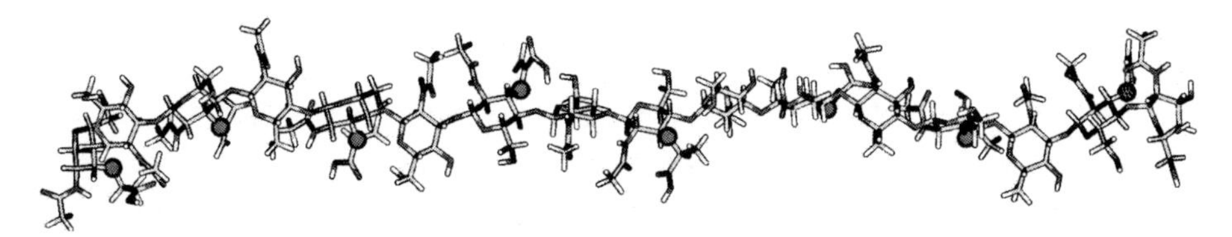

Figure 7.1 The conformation of a glycan of eight polymerized NAM-NAM disaccharides. This chain of 16 hexoses under no external stress forms itself into a helix spontaneously. In this chain, it takes four disaccharides to complete a revolution. This probably defines the conformation of the chains in bacterial walls and hence the two- dimensional structure of a layer of murein. It also leads to the tessera structure of the next chapter. The ether oxygen atoms in the D-lactyl groups of the NAM residues have been enlarged and darkened to make it easy to observe the roughly helical rotation of this relaxed glycan chain.

D-lactyl groups of the NAM residues have been enlarged and darkened to make it easy to observe the roughly helical rotation of this relaxed glycan chain. This is important confirmation of the earlier work of Labischinski *et al.* (1979, 1983, 1991; Leps *et al.*, 1984), which also implied that D-lactyl groups of the muramic residues rotate approximately 90° from disaccharide to disaccharide along the stress-free chain.

The magnitude of this rotation is very important for microbial physiology, since it means that the penta-muropeptide chains are only in a given plane at every fourth saccharide, and is only in the plane of the surface as well as pointing in the parallel direction every eighth saccharide unit. The conclusion is that the murein network cannot be very tightly knit and that the smallest possible structure covering an area will be eight saccharides long on each glycan side and two nona-muropeptides long on the other two sides. This unit of structure is called a "tessera". Moreover, it will not be a rectangle but closer to a hexagon because there will be two penta-muropeptides protruding from this structure in the middle of the glycan sides. These groups will extend away and be parts of other tesserae in the saccular wall. It is important that all of the tesserae will be subject to turgor stress, causing them to expand. More discussion of this is presented in the next chapter which is devoted to the unit of wall area, the tessera.

There is one other important implication of the structure shown in Figure 7.1 concerning the glycan part of cell walls. This is that one-half of the penta-muropeptides are pointing perpendicularly to the surface. Of these, one- half of these point inwards and one-half of them point outwards. In the case of the Gram-positive cell some of these may be linked to the adjacent layers, both to the inside and outside. But in the case of the Gram-negative cells the protruding muropeptides may not be able to be linked at any time and these free peptide chains may be subsequently degraded.

GLYCAN LENGTH DISTRIBUTION

The size distribution of the glycan chain lengths is relevant in discriminating between various models for wall growth in bacteria. To determine the distribution experimentally, the murein must be isolated, the peptide side chains cleaved off, and the lengths of the glycan chains determined by a chromatographic procedure. This was done first by Tipper (1969) for a Gram-positive organism and more recently and extensively in the laboratory of Höltje and Schwarz (Höltje and Schwarz, 1985; Harz, Burgdorf, and Höltje, 1990; Obermann and Höltje, 1994; Kraft, 1997). For *Escherichia coli*, Obermann and Höltje (1994) measured the length distribution in normal growing cells. Then, by subtracting the similar data obtained with minicells preparations; they estimated the chain length distribution in the cylinder part of the cell. The distributions of disaccharides for *E. coli* were observed to be continuous with a mode of about 79 disaccharides. The distribution is positively skewed, meaning that there are a few quite long ones. An important conclusion that they drew is that the glycan chains are too small to encircle the cell. Additionally, because they are not all the same length, consideration of certain modes of wall construction can be excluded.

Various models to fit the distribution of experimentally observed lengths were tried (Figure 7.2). It was found that it was sufficient to assume only two

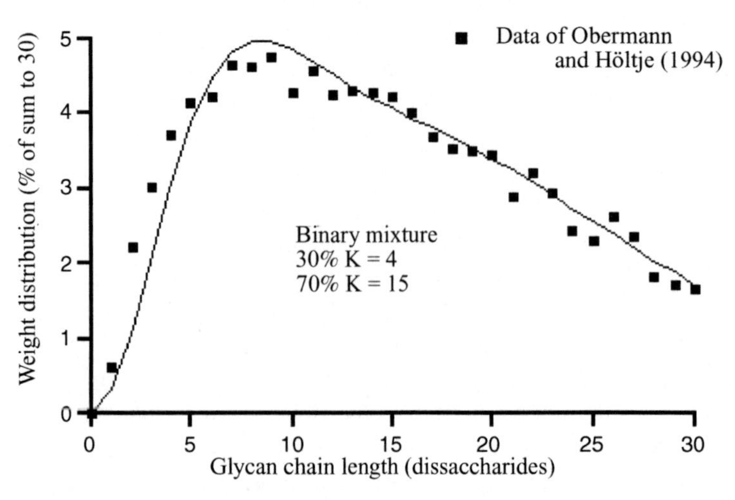

Figure 7.2 The distribution of the glycan chain lengths compared with the experiment. In the fitted distribution two components were assumed.

factors: first, adding and cleavage of disaccharide penta-muropeptides occurs at random in the growth and remodeling of glycan chains, and second, the assumption that the wall is less easily hydrolyzed in the axial direction than in the transverse direction. It is well known that the tension in a pressurized cylindrical wall is twice as great in the circumferential direction as in the axial one, so one class would consist of those chains aligned longitudinally, subject to lower stress, and would have a higher energy of activation for autolysis than chains aligned circumferentially. A good fit is obtained on the assumption that there are only these two classes of chains, with one more likely to be cleaved than the other. While further investigation is necessary, the key conclusion is that the length of the chains appears to be determined by a random process and not by a systematic mechanism that would generate a regular arrangement.

FORMATION OF THE NONA-MUROPEPTIDE

Biochemical studies have shown that the last stage of synthesis of the disaccharide penta-muropeptide is the attachment to the second saccharide, NAG. At this point the NAM is already attached to the penta-muropeptide, and is also attached to bactoprenol. As such, the disaccharide penta-muropeptide is in compact structure (Koch, 2000). The compact form would aid in its passes through the cytoplasmic membrane. Once passed through the cytoplasmic membrane and after it has entered the hydrophilic region (i.e., the periplasmic or outside region of the cell) it muropeptide unfolds, and the two functional protein-like groups, the diaminopimelic acid and the D-Alanyl-D-Alanine, are exposed (see Figures 6.2 and 6.3). When two muropeptides meet together with a trans-endopeptidase, a tail-to-tail bond can be formed. There are two possible pairs possible between the donor and acceptor penta-muropeptides. The important point is that, for physical reasons, only one bond can be formed. The remaining two groups will be attracted to each other by charge interaction, but for geometric constraints cannot form a covalent bond.

This geometry leads to the basic assumption of the "nona-muropeptide stress model" of bacterial wall growth; i.e., that this secondary charge attraction is important in protecting the adjacent covalent tail-to-tail linkage from enzymatic attack. It is also important because it initially occludes the two ionically interacting sites from chemical modification such as extension by linkage to an incoming penta-muropeptide.

The growth of the murein wall according to the nona-muropeptide stress model is as follows: after the primary linkage is formed, and as the cross bridge is stretched by growth, the nona-muropeptide will have its conformation altered as a consequence of cytoplasmic growth which forces murein wall area to stretch.

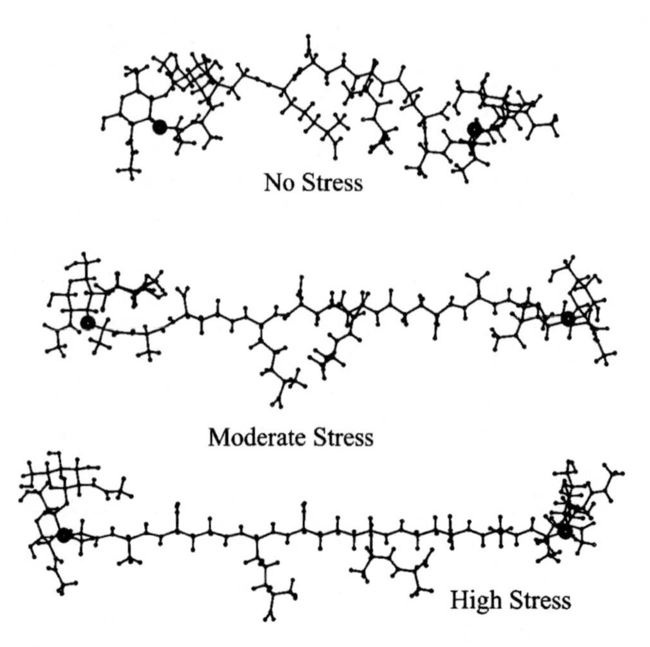

Figure 7.3 Nona-muropeptide at three degrees of stress. The conformation in an aqueous environment of a nona-muropeptide under various degrees of elongation is shown. The conformations from top to bottom are of low, intermediate, and high stress.

Consequently the conformation of the nascent peptide has to elongate. Computer-simulated conformations are shown in Figure 7.3 for three states of stress: relaxed, partially stretched, and highly stretched. Before the nona-muropeptide becomes stressed, the two unlinked groups interact with each other by charge interaction. Gradually as the chain elongates, these supernumerary groups tend to pull apart and lose their interaction. Then they can interact to form covalent bonds with incoming penta-muropeptides. See Figure 7.4 for the model relevant to Gram-negative cells, but see Figure 7.5 for a more generic model. These new groups may or may not be already linked to other peptidoglycan components.

A STRESS-CONTROLLED MODEL FOR BACTERIAL WALL GROWTH

The computer simulations of Figure 7.3 suggest a new possibility for a mechanism for safe wall growth. The model discussed here depends on the development of stress after insertion of new material into the wall. The structural geometric relationship is that an increase in tension in a nona-muropeptide causes the integral penta-muropeptides chains to elongate their conformation and cause

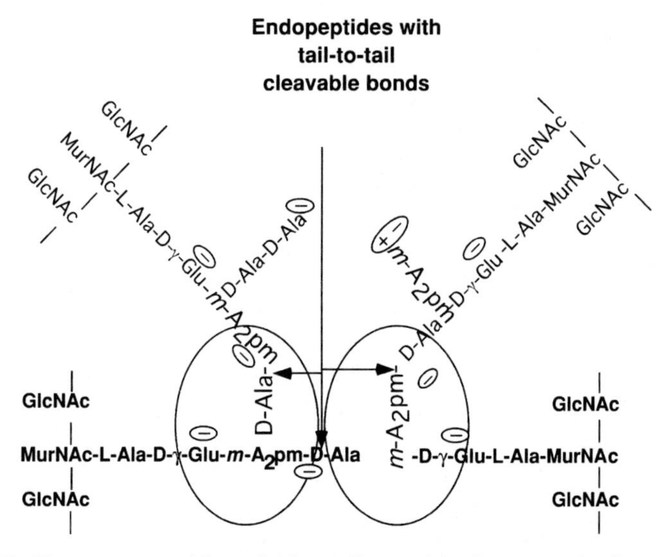

Figure 7.4 The Nona-muropeptide model for wall growth in Gram-negative bacteria. A Nona-muropeptide is shown at the bottom. Above it have been attached two penta-muropeptides. The ellipses represent the transpeptidase molecules that have linked them. A third arrow indicates where an autolysin can enter and cleave the old tail-to-tail bond that must be split to allow enlargement of the wall

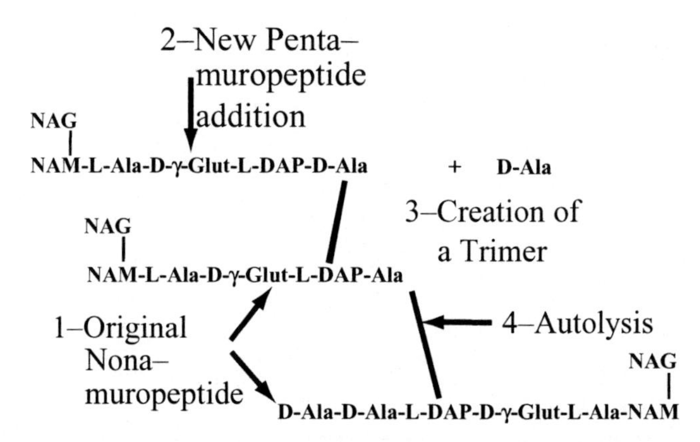

Figure 7.5 The generic model for wall enlargement. The steps in the enlargement start with: (1) a nona-muropeptide within the wall. It had been formed from two penta-muropeptides previously. (2) A new penta-muropeptide enters and is bonded by transpeptidation to a free group of the nona-muropeptide. The D-Ala that is liberated escapes. (3) The transpeptidation forms the trimer, which is composed of three muropeptides. The trimer is decomposed in the forth step by an autolysin to complete the enlargement by the safe incorporation of a penta-muropeptide.

the original ionic association of the two free tails to be broken and exposing them and their immediate environment to the outside environment. By exposing the underlying cross bridge, the existing tail-to-tail bond can be cleaved by an enzyme from the outside. In addition, this would allow two penta-muropeptides to enter the stressed wall structure and to be linked by their free m-A_2pm and D-Ala-D-Ala residues to form two new trimers. See Figures 7.4 and 7.5.

Thus cell growth and new insertions of muropeptides together with the resultant tension changes the conformation of a newly formed nona-muropeptide, and this aids and abets the Arrhenius principle in creating the opportunity for cleavage of the wall, permitting further expansion of wall in response to growth of cytoplasmic constituents. As shown in Figure 7.4, the opening of the ionic interaction leads to the possibilities of access of an endopeptidase to act on the existing critical tail-to-tail amide bond. This is besides the introduction of two disaccharide penta-muropeptides leading to the formation of two new trimers. Thus, the model proposes that this opportunity for growth of wall has been made possible and regulated by the tension in the nona-muropeptide chains in a way that is contingent on growth of cytoplasm.

There are two component chemical structures that are present in small amounts compared with the total murein of the bacterial wall; these are the nona-muropeptides and the trimers. The new model proposes that both are key intermediates for the wall growth process. Both are present in the small fraction of the total wall (>5%) because they are only present in the wall regions that are actually growing and not in the bulk of the wall that is not growing at any particular time. This idea of the importance of a component quickly formed and quickly destroyed is key for this model, as it is for the "three-for-one" model of Höltje (1993). He too argued that a growth intermediate, like an enzymatic activated state, would be expected to be present in small amounts.

Recently, it was found that the cylinder wall of *E. coli* grows in patches (De Pedro *et al.*, 1997, 2003 and unpublished observations). This is consistent with the idea that growth of the cylinder occurs at a given time only in small regions. It is suggested that new wall is created only in regions containing nona-muropeptides. Whereas in other regions (the completed poles) the wall is not growing because the nona-muropeptides have been depleted of one or both terminal D-ala residues to become octa- or hepta-muropeptides. In this way, one can explain how the wall grows in patches containing nona-muropeptide regions. If not too much patch area is present in the cylinder part of the rod-shaped gram-negative cell, then growth will continue as a cylinder. This has been shown by mathematical simulation (Koch, unpublished). An absence of nona-muropeptide could be also part of the reason that the poles of gram-negative cell are inert and do not enlarge their sacculus (see Koch, 2000 for relevant reference.).

In brief, the model proposed here is that, as the newly formed nona-muropeptide chain becomes extended under increasing stress stress, the unlinked *m*-A_2pm and D-Ala-D-Ala groups in a nona-muropeptide would separate to open a space for entry of new wall material to enlarge the wall. Wall enlargement could occur simply by inserting two disaccharide penta-muropeptides into the gap and forming trimer chains with these protruding residues. If these two penta-muropeptides have been linked (or became linked) to glycan chains, then the stressed underlying bond in the original nona-muropeptide could be selectively autolyzed by an endopeptidase and a new tessera created.

Chapter 8

The Structure of the Tessera; the Unit Structure of Murein Wall

To enclose the surface area of a bacterium, a very large number of the penta-muropeptides have to be covalently linked together. This fabric can either be considered to be a network or as a porous material containing a large number of pores. The smallest functional unit of surface around a pore, delimited by a ring of covalently linked atoms, is called a tessera. The covalent structure of the tessera has a rigid, strong structure resulting from its formation from ten disaccharide penta-muropeptides. There are two muropeptide pairs linked to each other to form two nona-muropeptides, two muropeptides point above the surface plane, two of them point below, and two point outward within the plane and are part of other tesserae. The tessera as part of a cell wall is a distorted square (approaching a hexagon) in which pairs of opposing sides consist of five linked disaccharides and in which one pair of sides is composed of cross-linked muropeptides. The glycan sides are distorted because in the growing cell all parts of the bacterial cell walls are under tension. The glycan chains are connected to other peptides at the middle disaccharides and stretching stress exerted there. A single tessera is a very small part of the total wall area; there are hundreds of thousands to millions of these tesserae in the wall of a bacterium. These pores in the wall allow the passage of small molecules in and out.

THE TESSERA

The strongest unit within the wall fabric is the smallest tessera. It has two deca-saccharides of alternating N-acetyl-glucosamine (NAG) and N-acetyl-muramic acid (NAM) residues on opposite sides (see Figures 8.1 and 8.2). Figure 8.1 is more diagrammatic than Figure 8.2, which shows the structure in more detail. The tessera has two glycan chains cross-linked by two nona-muropeptides linked through the NAM residues. The muropeptide shown in Figure 8.2 are characteristic of *Escherichia coli* but Figure 8.1 is generic and would apply to most bacteria.

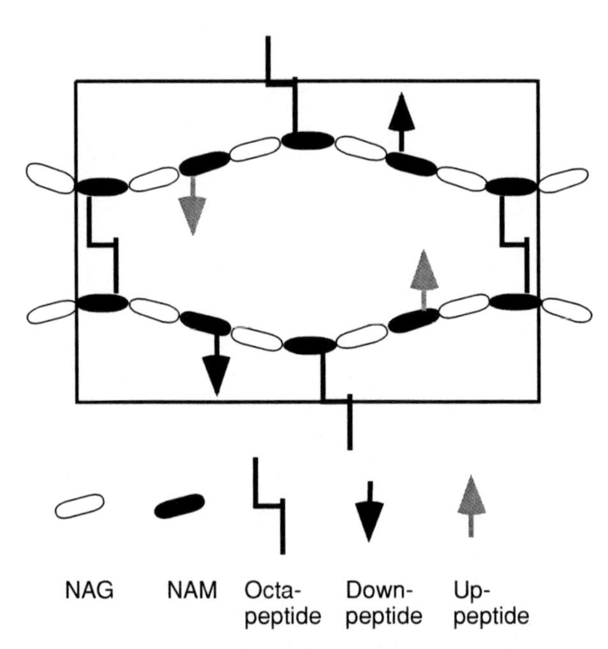

Figure 8.1 The Tessera. The glycan chains are shown as series of alternating filled and empty ellipses. The empty ones correspond to N-Acetyl-Muramic acid and have a carboxyl group that connects them to the muropeptide. Some of the muropeptides point up or down, but one-half of them are octa-muropeptides and link to another glycan chain in the wall.

The carbohydrate chain when free, with or without penta-muropeptides, and uncross-linked in dilute aqueous solution, would have its sugars and the peptides rotated about 90°, giving a full turn for eight sugar residues (see Figure 7.1). Consequently, the natural angles are not far from the ones needed to be part of a single-layered fabric. The angles could readily be distorted to create exact right angles between the orientations of the successive peptides emanating from a chain. In the depictions of the tessera, these are shown radiating away from the glycan chain. The peptides projecting up and down are indicated but not drawn out explicitly.

THE UNIT OF GRAM-NEGATIVE AND GRAM-POSITIVE WALL ARCHITECTURE: THE TESSERA

The results presented in this chapter have implications about the fundamental unit of the wall fabric. It is the protection to the cell given by the structural unit of wall that is the important subject at hand. Although the chemistry and biochemical studies in the past have of necessity focused on the linear chains of

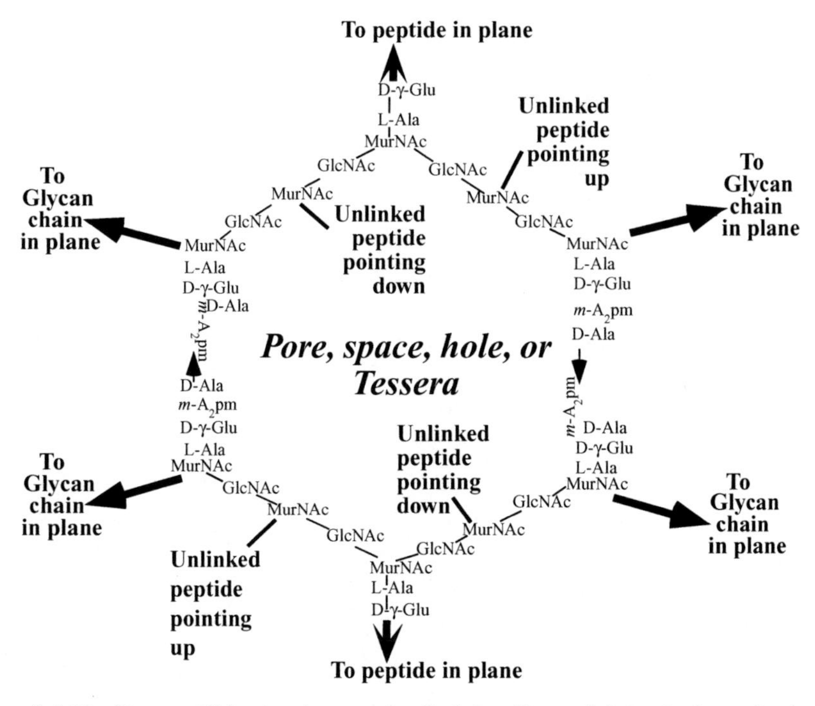

Figure 8.2 The Tessera. This view is more detailed than Figure 8.1, but both emphasize that the tessera is the framing around the pore that the cell uses for transport materials into and out of the cell. Notice that a different set of abbreviations is used than was used above with MurNAc for NAM and GlcNAc for NAG.

sugars or amino acids, of course, the polymer of either the cross-linked peptides or glycan chains have to lie in the plane of the cell surface. The entire coverage by the fabric covering the surface area would be important to the cell. We have defined the smallest indivisible unit of surface area to compose the unit of the bacterial wall mosaic as a tessera. The tessera is the two-dimensional equivalent of the unit cell of crystallography. A complete collection of tesserae bound together is the geometric entity that could fully form a complete single-layered surface. It would be a curved two-dimensional surface that would cover and enclose the cell. The covalently bonded structure of atoms that corresponds to the physical edges of the tesserae is a two-dimensional network; i.e., it is a ring consisting of both sugars and amino acid elements. Each chemical grouping is part of multiple tesserae. (From the connectivity of the fabric, conceive of a physical model by picturing a fence of chicken wire that has hexagonal holes. In that case, a segment of wire defines the limits of the individual structural porous units, while in the bacterial case, it is two cross-linked muropeptides and two portions of glycan chain that do so.)

THE *IN VIVO* SACCULUS IS NOT MAXIMALLY EXTENDED

Living *E. coli* cells shrink 45% in surface area when the turgor pressure is eliminated by detergent treatment; they shrink 20% when the osmotic pressure is raised to a point assumed to erase the turgor pressure difference. The finding that the relaxed isolated sacculus can readily expand 300% in surface area (Koch, 1985; Koch *et al.*, 1987) with quite minimal stress forces seems contradictory to these findings. There are two interpretations of these results. The first is that the inner and outer membrane contributes a good deal to supporting the stress due to turgor and to keep the murein from expanding to its fullest extent. The second (much less likely) interpretation is that the turgor pressure does not present as strong a stress as the electrostatic repulsion that we applied by causing the sacculi to bear a net positive or negative charge in making these measurements.

The second unexpected result of the studies reviewed in this chapter is the key role of the zwitter group of diaminopimelic acid. When the carboxyl and amino groups are very close to each other due to their connection to the same carbon atom, they cause the pK of each to be more extreme. The finding that saccular expansion is associated with either pH extreme (Koch and Woeste, 1992) shows that this part of the sacculus is key to its expansion whenever and however the zwitter character is destroyed and becomes a group that has

Table 8.1 Mean sacculus surface area after various treatments[a]

BHI grown sacculi[b]		BHI grown whole cells[c]	
	μm^2		μm^2
pH 4.8 1 M KCl	6.1	Agar filtered OsO4	6.9
pH 1.5 H2O	22.5	Agar filtered Glut	8.1
pH 1.5 Urea	24.5	Phase Contrast	8.6
pH 12.5 Urea	20.4		
pH 6.0 Acetylated	21.5		
pH 6.0 Succinylated	25.9		

[a] The mean sacculi surface area measurements from the light scattering measurements of Koch and Woeste (1992) and the mean estimates of the surface area of bacteria from the electron and phase microscopic data of Woldringh (Zaritsky *et al.*, 1979). All refer to experiments with strain B/r H266 grown in brain heart infusion (BHI) media at 37°C with aeration.

[b] Growing cells were harvested, lysed with SDS, purified with trypsin, washed extensively, and analyzed by the low-angle light-scattering method developed in our laboratory (Koch and Woeste, 1992).

[c] Data from Conrad Woldringh for agar filtered cells fixed either with osmium tetroxide or glutaraldehyde and by phase contrast microscopy in the living state (Woldringh, 1985).

a single charge of either sign. This finding may be particularly important in considering the implications of elasticity or of inelasticity and lack of ability to expand in comparing the shapes of the poles of *E. coli, B. subtilis*, and of *E. hirae*.

Table 8.1 shows experiments with *E. coli* sacculi which were treated in various ways to cause them to stretch or contract.

Chapter 9

Extrusion and Incorporation into Wall

After the basic disaccharide penta-muropeptide unit has been prefabricated, it is delivered to the building site. The problem is that this site is on the other side of an obstacle; i.e., the cytoplasmic membrane. Lipid-soluble compounds can pass through the bilayer, while polar molecules of the size of this basic unit (molecular weight ca. 1000) cannot. The bacterial strategy for the export of this wall unit is that the penta-muropeptide is attached to a non-polar membrane-bound carrier, bactoprenol. Subsequently the hydrophobic bactoprenol molecule (while remaining in the membrane) functions to aid the penetration of the disaccharide penta-muropeptide through the membrane. Finally the unit is dissociated from this "carrier" and attached; i.e., inserted between another carrier and the growing glycan chain that may be composed of a number of peptidoglycan units. The insertion of glycan units and the cross-linking of the peptides is a complex biochemical and the biophysical process that is quite complicated, but is not well understood. Some of the larger penicillin-binding proteins (PBPs) have transglycosylase activity as well as transpeptidase activity; they are, no doubt, involved at this stage in forming both linkages. Besides glycan formation, cross-linking of two muropeptides units occurs by transpeptidase action in the formation of a tail-to-tail linkage.

THE CHEMICAL NATURE OF THE PENTA-MUROPEPTIDE-BACTOPRENOL COMPLEX

The unit structure of disaccharide penta-muropeptide that has to pass through the wall was shown above in Figure 6.1. This figure is misleading because that structure was stressed in the computer to straighten the molecule out. Because of hydrophilic nature of the aqueous periplasmic environment this is closer to the natural conformation in the murein. Conversely, in order to model the penta-muropeptide for its conformation in an environment approaching that of the phospholipid membrane, it was necessary to set the dielectric constant to 1 instead of the value characteristic of a watery environment where the value would be approximately 80. In this way Figure 9.1 was obtained.

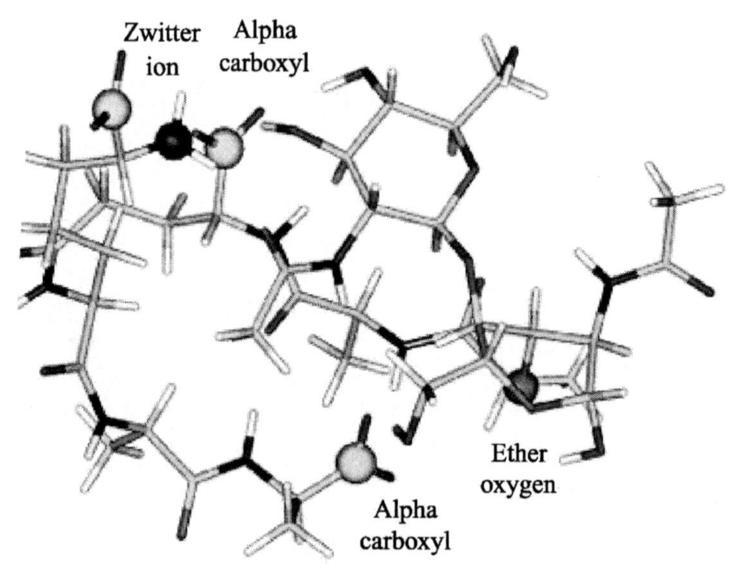

Figure 9.1 A murein unit in the conformation suitable for transport through the membrane. A computer simulation (Koch, 2000) is shown of a penta-muropeptide in an environment simulating the hydrophobic environment of the cytoplasmic membrane. The important point about this structure is that the adopted conformation brings the zwitter ionic group at the end of the diaminopimelic acid to be positioned closely to the α-carboxyl group of the D-glutamic acid of the penta-muropeptide chain. Energy minimization led to a conformation with the amino group located between the two-carboxyl groups and thus the hydrophilic character of these ions is partially lost. This would help the penta-muropeptide to expose a more hydrophobic exterior as it is passed through the phospholipid membrane. In this figure the charged groups are emphasized and marked by the balls on the carbon atoms attached to the groups that have net charges. In addition a ball is located on the one oxygen atom that is in ether linkage. This atom connects the disaccharide to the penta-muropeptide chain in the wall unit.

INSERTION OF A NEW UNIT INTO A GROWING CHAIN

Once the penta-muropeptide, attached through a pyrophosphate to the bactoprenol is flipped to the outside of the cytoplasmic membrane, the problem is how to incorporate it into a growing glycan chain. Even if a new chain is to be created the same problem must be faced. Some of the energy for linking together the new disaccharide to a growing glycan chain is probably provided by the pyrophosphate, which becomes hydrolyzed by a pyrophosphatase as the disaccharide penta-muropeptide is transferred into the growing glycan chain. But the enzymatic problem remains that the unit's linkage to the bactoprenol must be split and the disaccharide muropeptide has to be added to another wall unit. From a general knowledge of biochemistry, it might be expected that the phosphate

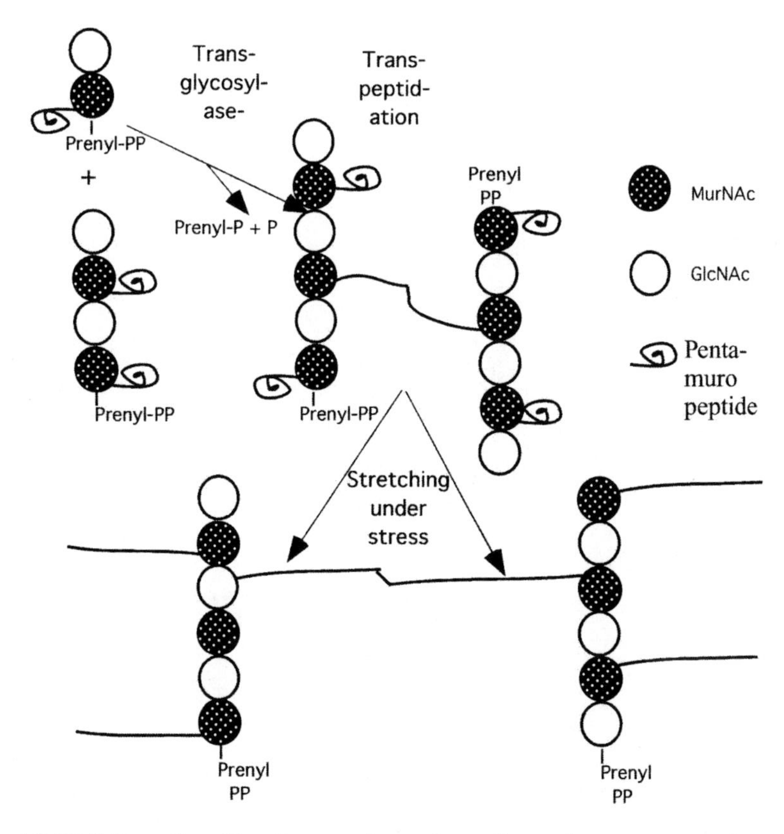

Figure 9.2 Wall formation. The diagram shows the wall growing process as viewed from above the cytoplasmic membrane. On the upper left, the bactoprenol-linked disaccharide penta-muropeptide is shown being attached to the terminal region of a growing chain. Hydrolysis of the pyrophosphate linkage supplies the energy for the formation of the glycosyl bond and drives it to completion. The prenyl-P turns around to face the other side of the membrane in order to engage in forming another disaccharide penta-muropeptide. The liberated phosphate is pumped into the cell. The growing chains form endopeptide bonds as shown on the upper right side. At this point the nona-muropeptide is in a relaxed conformation, but as the cell grows the murein is stretched and stressed.

group would be replaced by the hydroxyl group from the other disaccharide in a spontaneous reaction with a negative free energy. This reaction might proceed because the negative free energy of the phosphate-sugar bond is larger than that of the glycoside bond that is then formed between two sugars.

This chemistry is shown in Figure 9.2 on the assumption that the new unit is added to the older part of the growing chain. For this assumption there is no experimental evidence at present. It is also to be noted that the structures shown in the figure have to lie flat on the cytoplasmic membrane because this is the locus of the lipid-soluble bactoprenol molecules.

Chapter 10

The Role of Poles in the Growth Strategy of Bacteria

As the additional cytoplasm is formed, it must be encased by a larger cell wall, and therefore, wall growth is imperative. For both cocci and rods, the extant poles are rigid and inert and serve as geometric support for cell expansion. For rods, much of the cell growth is accommodated by cylindrical elongation. Critically, the existing poles are the support for maintaining the cylindrical elongation at a constant radius. For cocci the only means of enlargement is division; i.e., the formation of a septum and its splitting and bulging, which also depends on the extant poles having a constant radius. This process leads to the new poles for the two daughters and to the accommodation of more cytoplasm. Just how the metabolism of various parts of the wall is controlled is important, but not fully clear at present.

HOW DO BACTERIA KEEP FROM GROWING EVER LARGER?

Bacteria do not ordinarily grow bigger and bigger and rounder and rounder. A logical possibility to explain this constraint is that the poles of bacterial cells are rigid, metabolically inert, and cannot stretch or deform. This could act as the framework for the structure of bacterial cells that have rigid, strong walls. Such an explanation was given two decades ago based on evidence suggesting that the poles of a Gram-positive coccus and a Gram-positive rod were metabolically inert. Today, there is critical evidence for the inertness of the poles of *B. subtilis* and *E. coli*. Although the poles of many bacteria are quite inert and hardly turn over, the sidewalls of many rod-shaped cells have a metabolic half- time of about the same length as their doubling time. The inertness of poles is presumed to be necessary to support cylindrical growth and also support new pole formation. The fact that the poles remain of constant diameter from generation to generation (but not always in *E. coli*, see below) has led to a general theory for bacterial morphology, i.e., the surface stress theory. However, an understanding of the mechanism for inertness in different regions of the same cell has not been established.

Two hypotheses for metabolic inertness of poles have been presented. Both are reasonable, but critical evidence has not been presented for either. For Gram-positive cells, the hypothesis is that the orientation of the glycan muropeptide chains in the pole wall is different from their orientation in the sidewall so that the poles may be resistant to being attacked by autolysins. For Gram-negative poles, the thin wall may be unable to grow in regions where the cross-linked nona-muropeptides have had the remaining terminal D-alanine removed. The hypothesis is that the murein in the newly formed poles loses its unbound terminal D-alanine by the action of carboxypeptidases. It is possible that penultimate D-alanine may be also cleaved off by carboxypeptidases.

The surface stress theory has been useful but has always been weakened, until recently, by a lack of understanding of how the poles and most of the sidewalls can differ metabolically. This is even though the need for pole rigidity by the cell can be supported by many arguments. The only hypotheses available accounting for the inertness and lack of metabolism in different regions of the Gram-positive and Gram-negative bacteria, are presented in this chapter.

AN EXPLANATION FOR INERTNESS OF THE POLES OF GRAM-POSITIVE RODS

A possible explanation for the lack of turnover of the Gram-positive pole involves the conversion of the planar septum into two nearly hemispherical poles. This geometric conversion leads to the suggestion that autolysin's action is more effective in acting at right angles to a plane of a murein layer than within the plane of that layer. This follows from the geometry of the binding of the active site of the autolysins with sensitive bonds in the murein.

In analogy, the autolysins act better "across the grain" than "into the grain". From Figure 10.1 it can be seen that the splitting of the septum to form the two poles occurs at right angles to the surface of the sidewall layer, whereas any autolytic attack on the polar surface would be parallel to the plane of layers of the murein. This would result in much more rapid attack at the sidewall than at the poles.

AN EXPLANATION FOR INERTNESS OF THE POLES OF GRAM-NEGATIVE RODS

The above explanation would not work for the murein monolayer present in the Gram-negative poles and sidewalls. A different possibility, at least for

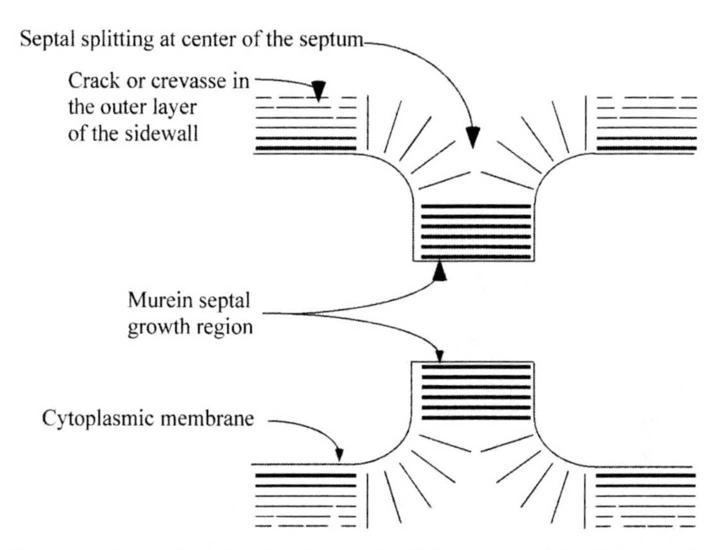

Figure 10.1 A potential explanation for the metabolic inertness of the Gram-positive septum and pole. The inside-to-outside growth mechanism for the sidewalls, discussed in more detail in Chapter 12, is diagrammed. The wall is laid down in layers. Subsequent layers push the earlier ones outward and they become elongated as the cell grows. Eventually they cannot stretch any more and are then split and discarded by an autolytic process.

For the cell division, a planar septum forms, growing inward by the addition of new wall in successive layers. For a coccus the septum is initiated by splitting the cell into two equivalent poles; for the rod, the split occurs in the middle of the cylindrical region. Septal growth is as in the inside-to-outside sidewall growth process, but in the opposite direction. This is shown in the central part of the figure. This wall is compact, dense, and unstressed until a central autolytic split bisects and exposes the septal wall to stress. The surfaces produced are normal to a bisecting autolytic cut and thus different than those of the sidewall, and this can be expected to prevent autolysins that are present in the environment from acting and thus preserve the integrity of the pole.

E. coli, follows from a new model called the "nona-muropeptide stretch model". This hypothesis arises from new computer modeling studies of the conformation of cross-linked peptidoglycan shown in various parts of this book and from the experimental and modeling studies of De Pedro *et al.* (1997, 2003) of the pattern of insertion of new murein into regions of the cell wall (see Chapters 6 and 7).

In the linking of two muropeptides to form the cross linkage holding the wall fabric together, a D-Ala-D-Ala of one muropeptide reacts with the amino group of the diaminopimelic acid of an another group in a transpeptidation reaction that binds the two muropeptides in a "tail-to-tail" linkage and liberates a molecule of D-alanine. After such a linkage is formed, there still is a second D-Ala-D-Ala group and a second diaminopimelic acid group in the other orientation; both of these could react these with additional muropeptides in the same way and thus create two trimers (see Chapter 7).

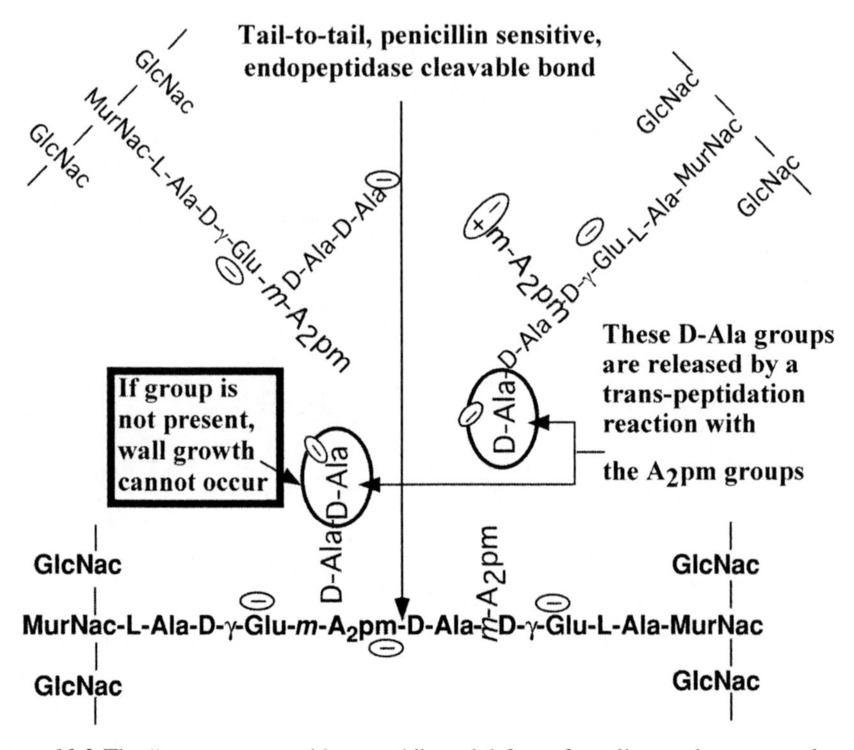

Figure 10.2 The "nona-muropeptide stretch" model for safe wall growth assumes that the unlinked D-Ala-D-Ala and diaminopimelic acid groups of newly formed nona-muropeptides electrostatically interact after the cross bridge is initially formed, but that as the result of sufficient stress during further growth, this ionic bond would open exposing both the free D-Ala-D-Ala and free diaminopimelic acid groups of the nona-muropeptide. These then react with new disaccharide penta-muropeptides to create two new trimers and also open a space to permit the splitting of the old nona-muropeptide, thus allowing the wall to elongate.

However, this enlargement can only occur if the second D-alanine had been retained by the original cross-linked muropeptide; i.e., if the cross bridge is in the nona-muropeptide state and has not yet lost its terminal D-Ala by hydrolysis. The consequence of this situation is that wall growth can only occur where the D-Ala-D-Ala group remains on the cross-linked muropeptides and that wall enlargement is blocked where this group has been hydrolyzed to leave a single D-Ala. This model postulates that it is this removal of the D-alanine groups that renders the pole wall inert to further metabolism.

For wall growth in the model presented in Chapter 7, two trimers need to be formed on a cross bridge and then the original tail-to-tail bond needs to be cleaved. While a tail-to-tail bond can form when the incoming group provides the D-Ala-D-Ala and the tetra part of the penta-tetra cross-link that constitutes the nona-muropeptide donates its diaminopimelic acid group, if the penta part has lost its D-ala (to become a tetra) then only one trimer can form see Figure 10.2. The model proposes that the murein can enlarge only when two trimers are formed and the old one is resolved by the cleavage of the original cross bridge.

Without the presence of a D-Ala-D-Ala peptide bond on the existing nona-muropeptide cross bridge, insertion and enlargement by transpeptidation of a new muropeptide unit cannot occur. This implies that hydrolysis of the free D-Ala-D-Ala on the stress-bearing wall by one of the carboxypeptidases, with the liberation a D-Ala residue and creation of an octa- or hepta-muropeptide, blocks local wall growth because the peptidoglycan cannot enlarge. This is what is assumed to occur in polar wall regions of the cell (and in patchy areas of the sidewall); this renders the poles and patches on the sidewalls inert.

This explanation was developed from the computer simulation of murein structure and is in accord with the recent results of De Pedro *et al.* (1997, 2003) and my further analysis of the fluorescent microphotographs of cells from their experiments. Figure 10.3 shows another fluorescent microphotograph obtained with his technique but in a different computer representation and of a different cell than shown above. De Pedro's method follows the temporal formation and insertion of murein into the wall of growing bacteria by the incorporation of D-cysteine. Growth in the presence of this abnormal amino acid causes the wall to have about 5% of its D-alanine groups replaced with D-cysteine. The

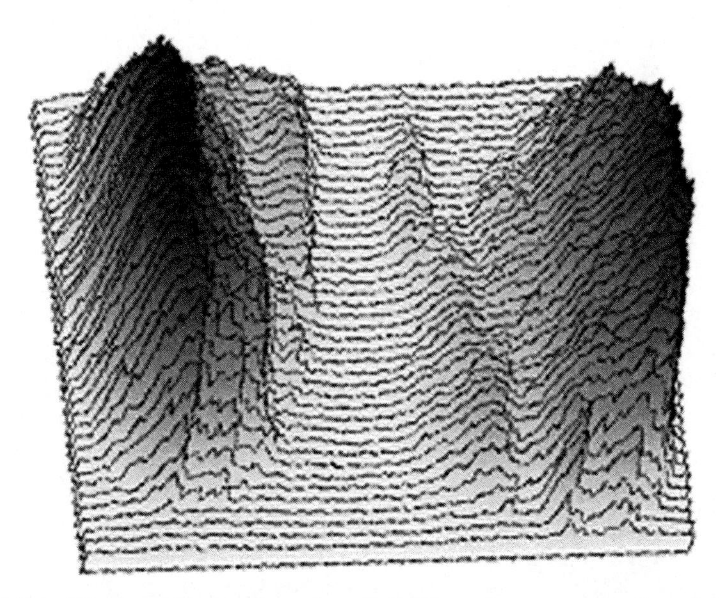

Figure 10.3 A different representation of a cell dividing after one generation of chase. The representation of the older murein present in the sacculus is at a higher elevation in the figure. It can also be seen that the photographic saturation process itself decreased the amount of old murein at the poles. If this had not occurred the elevation would be higher than shown.

substitution of D-cysteine for the D-alanine depends on an amino acid exchange reaction taking place in the periplasm. The D-cysteine content of the wall in different regions (pole, sidewall, and division site) of the cell permits the pattern of temporal insertions into the cell wall to be deduced. The D-cysteine levels in the wall were followed by staining with a fluorescent antibody.

By labeling and then chasing, it was shown that extant poles were not significantly diluted with new material during further growth, while the sidewalls were a non-uniform mixture of new and old material, but the wall of the constricting division sites was made entirely of new material. Although earlier work had suggested the conclusion that the poles were inert and did not turn over, this new work is the definitive proof that the poles of the cells are inert. The new work also extended the findings of Schwarz *et al.* (1985) and Woldringh *et al.* (1985) and conclusively shows that the region destined for new pole formation was made entirely of new murein and not even a slight admixture of old with the new murein.

This new model for wall growth of a monolayered murein as shown in Figure 10.2, would permit the growth only if the D-Ala-D-Ala grouping in the wall remained intact. The new model requires one or more of the carboxypeptidases to be the basis of both the pole inertness and the patchy development of the sidewall. *E. coli* has several carboxypeptidases with the capability of splitting the D-Ala-D-Ala grouping and would in time convert the nona-muropeptides to octa- or hepta-muropeptides so that no transpeptidase enlargement of the wall in the pole, or any patchy regions of the sidewall, would occur; this would prevent further growth from occurring in those areas.

CONCLUDING REMARKS

An important part of the growth strategy of bacterial cells is that some cellular dimensions are constant during the cell division process. For example, *B. subtilis* cells have a diameter that is invariant to the richness of the growth medium, although the length at division is longer in the richer medium. Even though the size in *E. coli* changes, the ratio of length to diameter is quite constant, and the diameter is quite constant during steady state growth in a given medium. Size invariances may be important for the growth of all walled bacteria. The cellular architecture requires a fixed dimension of the poles of rod-shaped cells in order to form a framework for growth of the sidewall during vegetative growth. It is now established experimentally that the poles of many organisms (rods and cocci) are metabolically inert and do not turn over. From these experiments have arisen the theoretical and speculative ideas that I have presented in this chapter. They provide reasonable explanations for the rigidity necessary to permit a workable strategy for safe growth and maintenance of the shapes of most bacteria species.

An aspect that was critical for the development of these ideas was the existence of rod-shaped bacteria, the subject of the next chapter. The modeling studies with soap bubble pipes and soap solution lead to the idea that rigid ends are enough to allow cylinder segments of bacteria to elongate (see Figure 11.1).

Part 3

Bacterial Morphologies

Chapter 11

Sidewalls of Gram-Negative Rod-Shaped Bacteria

The wall of Gram-negative rod-shaped bacteria can be divided into three regions: poles, sidewall, and forming poles. The completed poles are inert, they do not turn over, and they function to mechanically support the elongation of the sidewall at constant diameter. The subject of the present chapter is to describe an analysis by labeling and staining of old murein with a method that does not stain the new and, by the use of computer methods, to analyze the images. The murein of cells that had been labeled with D*-Cysteine and then further grown (chased) in the presence of an antibiotic that blocks cell division has only recently become possible. In these studies, sacculi were prepared and visualized with a fluorescent antibody. One critical finding is that the central septating or constricting part is entirely new. This is not surprising because it was known that the septa had to be, at least, largely new. However, another finding is quite surprising. Although the cylindrical sidewall contains both old and new murein, it is a mixture of patches of old and of new material instead of a uniform mixture. These patches are hundreds of oligoglycan chains across and long and many nona-muropeptides wide. In a one-doubling time chase, the patches of new wall do not form bands normal to the cell axis as in the septal regions. On the other hand, they are not helical and thus they are not consistent with recent work with B. subtilis and Caulobacter crecentus. In a chase that lasts two-doubling times, there are three separate regions of new (fully non-labeled) material inserted in the sidewall. These substructures go perpendicularly entirely around the cell and are, probably, the beginnings of the next two cycles of division sites. There are indications of further cycles. In toto, these findings are inconsistent with all the different models in the literature about how the sidewalls and poles of the Gram-negative rods grow and divide and of the kinetics of the development of the division sites.*

SOAP BUBBLE ANALOGY TO BACTERIAL GROWTH

Figure 11.1 shows an experiment with a jar, a wire ring, and a soap solution that demonstrated that cylindrical soap bubbles could form under the

Figure 11.1 Demonstration that physical forces can generate cylindrical structures analogous to rod-shaped bacteria.

right physical conditions. This suggested that this could be the basis for the growth of rod-shaped bacteria.

STAINING OF THE OLD MUREIN

A few years ago De Pedro devised a way to follow the temporal formation and insertion of murein into the wall of growing bacteria by the incorporation of D-Cysteine (De Pedro *et al.*, 1997). Growth in the presence of this abnormal amino acid causes the wall that has been laid down to exchange the D-Cysteine for the terminal D-Alanine of the muropeptides under the aegis of a periplasmic enzyme. This results in about 5% of the terminal D-Ala groups being replaced with D-CYS. The D-Cysteine content of the oligopeptidoglycan chains in different regions (pole, sidewall, division site) of the sacculi permitted the pattern of temporal insertions into the cell wall to be deduced with a fluorescent antibody. By labeling and then chasing, De Pedro *et al.* (1997) showed that extant poles were not appreciably diluted with new material during further growth, that the sidewalls were a mixture of new and old material, and that the wall of the division sites was made almost exclusively of new material. This work was definitive proof that the poles of the cells were inert.

An unexpected aspect of the analysis of published photographs was that the sidewalls do not enlarge by random insertion of single disaccharide penta-muropeptides into oligopeptidoglycan chains distributed uniformly over the entire cylindrical region or according to a regular non-random pattern. It occurs in an irregular aggregated fashion in which the insertions are comprised of groups of many disaccharide penta-muropeptide units. These groups are wider and much longer than the observed length of the average length of 14 glycan residues per chain (Obermann and Höltje, 1994; Ishidate *et al.*, 1998).

Figure 11.2 shows a cell chased for one generation in the presence of aztreonam. This is a β-lactam antibiotic that block's cell division. This graph is a three-dimensional representation of the presence of the older material

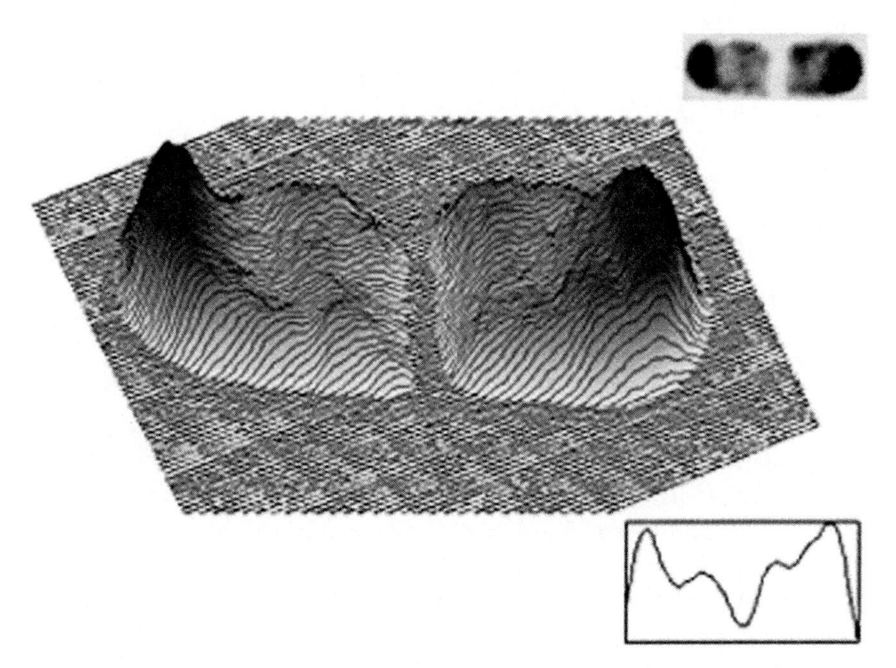

Figure 11.2 Several views of the age distribution of a cell chased for one doubling. The large image is equivalent to the model shown in Figure 11.1 in that it has a cylindrical region of the two cells produced by division. The upper right directly shows the fluorescent image and the lower right image shows the fluorescent density of a longitudinal section.

Figure 11.3 Evidence that the age of wall sections varies throughout the cell and that the pole regions are the oldest. This is a two-generation chase of a D-Cysteine labeled bacterium in the presence of aztreonam to block cell division. This figure shows that the sidewall after this length of chase is composed of largely old segments interspersed with new wall. Although it appears that this fluorescent micrograph is two cells and not one, other representations showed that the entire body of this cell is only one cell.

(old murein is shown in black). Several new features can be seen. Firstly, not only are the sites where division would have taken place in the absence of aztreonam, evident, secondly, possible starts for the septal regions for the generations yet to come are evident as well. Thirdly, while the poles appear to consist of only old material, the residual sidewall has patches of old material. There is more in regions that are nearer the poles. Figure 11.3 shows a two doubling time chase.

The novel observation presented in both figures is the irregularity, or unevenness, or patchy insertion of new wall in the cylindrical sidewall region. In the cylindrical parts of the cell not appearing to be part of the cell division regions, there are significant irregularities in density that suggest that the murein of the sidewall is not formed in a systematic, regular way. If the new wall were intimately mixed with the old in the sidewall region, then the density should be uniform but not as great as it is in the poles. This means that insertion of new wall does not occur at regular intervals within the old muropeptides and that it does not occur by the insertion, at random, of individual disaccharide penta-muropeptides into the stress-bearing wall or insertion of previously prepared groups of three muropeptidoglycan chains in replacing an existing chain as in the "three-for-one" model of Höltje (see below).

With a longer chase in the presence of aztreonam, the cell filaments grew longer and showed many more additional growth sites. These appear in Figure 11.1 to be represented in white, while the old murein appears dark. There appears to be 15–18 bands or hoops of at least mostly new wall going around the cell. This pattern was seen by computer analysis in many of the sacculi in the original microphotograph. These had the appearance of hoops or bands circumnavigating the cell, most at nearly right angles to the cell axis. Some but not all of these may be equivalent to where cell division might have, or would have, taken place if the aztreonam had not been used. Such bands are apparent in the other samples of two-generation chased filaments. This result is in contrast to the figures derived from the one-doubling time chase where the partially complete division sites were not fully perpendicular to the cell axis; in the longer chase many of the new bands, apparently completed, appear to be normal to the cell axis. Typically, there are three complete and possibly several initiating division sites.

During wall growth in the presence of the aztreonam, two types of wall growth occur. In sites where the cells would normally attempt division, a band or swath of entirely new wall is inserted between the existing sidewall without any constriction of the cell diameter. While this supports an idea in the literature for a long time, the present data is probably more critical than earlier studies. In the chase periods, this accounts for the clear region in the central parts of the sacculi made of entirely new murein and with very little or no older D-Cysteine-containing murein. While the conclusion that the cell division sites are made of new murein is not novel, the maximum amount of old murein intermixed with new is much smaller than could be established by previous methods. One could further interpret the lighter regions at the 1/4 and 3/4 positions in the one-generation chase as the beginnings of the next attempts at division. Many workers have suggested the concept that new sites start before the developing poles are completed. The new finding is clear, definitive evidence that they do and that they start even earlier than was thought. However, initially they are not regular and are not formed as hoops or bands perpendicular to the axis of the cell, but they are in the appropriate place for the next cell division.

MODELS FOR SIDEWALL GROWTH

These results bear on the mode of gram-negative wall growth. There are three kinds of models currently proposed to explain the phenomenon of cylindrical growth of bacteria; these are summarized in Koch (1998). The surface stress theory (SST) (Koch, 1983, 1988, 1991, 1995, 1998, 2000, 2001) explains the extension of rod-shaped cells by physical forces akin to surface tension

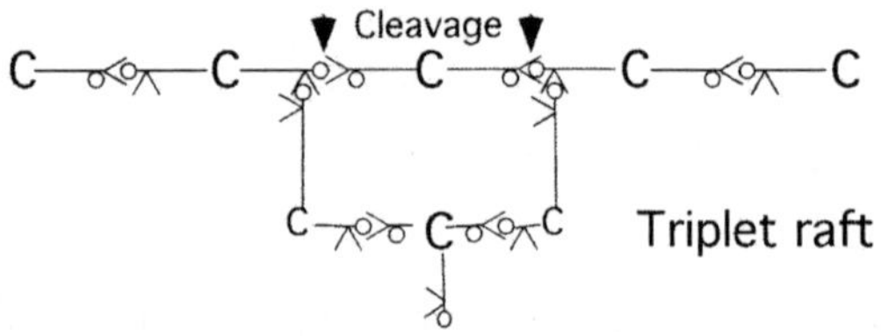

Figure 11.4 The three-for-one model of Höltje. The model proposes that a number of enzymes function together as a holoenzyme to simultaneously both attach new murein to the wall and to cleave the wall in such a fashion that the new material is pulled into the stress-bearing wall to enlarge it (Höltje, 1993). A key point to the model is that a chain of three penta-muropeptides wide and as long as the template strand on a stress-bearing glycan chain are bound to each other and this triplet raft is bound to the template glycan strand in the stress-bearing wall and, as shown, to enclose this glycan chain that is then cleaved out, replacing one chain with three of the same length.

acting, when the wall grows, as new units are inserted and then stretched in a plastic fashion within the intact wall of a growing cell. Because of its plasticity the sidewall is a non-rigid structure; however, it can maintain a constant diameter under the appropriate physical and metabolic restrictions. The key condition is that the poles are rigid and support the cylinder region as it grows (Koch *et al.*, 1981; Koch, 1983).

Competing models fall into two classes. The Höltje model (1993) is the most accepted model at present. It is outlined in Figure 11.4. Growth occurs by inserting a raft of three chains of murein in place of one template strand. It assumes that the linked chains are non-elastic and are connected to surround the circumference of the cell and prevent the wall from stretching. In order to prevent the chains from bulging, they would have to be under tension from the time of their formation. (There is an alternate version that could apply if the enlargement of the wall depended on a mechanism that could copy a template with a defined number of hexose residues.)

There is a third class of models which assumes the existence of mechanoproteins that could exert forces, causing extension over the length of the cell or contraction over the width of a cell (Norris *et al.*, 1994). This is in spite of the fact that there is little evidence of force-generating proteins in bacteria (Koch, 1991, 1998). Although clearly FtsZ and FtsA have a homology to tubulin and actin and are involved in cell division, they cannot have a role in constraining the diameter of the cell because of their known numbers of molecules and their location within the cell.

Now that the old poles have been critically shown to be metabolically inactive, and therefore can form a rigid structure on which the elongating sidewall and the developing division site can develop, it has become convincingly established that it is the completed poles that have the role of determining the radius of the cylindrical cell. In the De Pedro *et al.* (1997, 2003) experiments, another new finding was reported. It is that the region of the cell that would have been part of a septal constriction, if the applied antibiotic had not inhibited the PBP 3 function, remains at the same diameter as the poles even though the wall that is laid down is made entirely of new murein. It is interesting that the diameter of the cell remains the same as the sidewall, even though the regulation, enzymology and biosynthesis are quite different in different parts of the cell.

IS THERE HELICAL WALL IN *E. COLI?*

By studying multicellular structures, the concept that wall growth of *B. subtilis* occurs by helical insertion of murein was formulated by Mendelson in 1976. Later work (Koch, 1989, 1990) showed that a *B. subtilis* filament did indeed rotate as it grew, but this was due instead to the way the wall on the outer surface was autolyzed. More recently, workers in Errington's laboratory (Jones *et al.*, 2001) presented critical evidence that a closed helical structure formed in this organism in the absence of a cytoskeleton. Subcellular localization of the MreB and Mbl proteins revealed that each forms a distinct kind of filamentous helical structure lying close to the cell surface in the cytoplasm. If these structures are important in the growth and division of *B. subtilis*, then a closed helical structure must become converted into two helical structures as part of the division processes.

I have careful inspected the *E. coli* fluorescent micrographs from De Pedro's laboratory to see if the new wall formed during the chases could possibly be part of helixes, but I can see no evidence that this is so. Unpublished findings from Gober's laboratory (see England and Gober, (2001) showed helical structures to form under the wall of *C. cresentus*. So at present we must leave this question unanswered.

In summary: the findings that the newly forming (nascent or pre-septal) poles are made of entirely new material and that no old material is inserted into old poles require re-evaluation of the older ideas about wall growth and division. Additionally, re-evaluation is also required because the sidewall is a mosaic of areas of new and old patches and not an intimate, regular admixture of new and old muropeptides.

Chapter 12
Growth Strategies for Gram-Positive Cells

It has been known for more than a century that some bacteria retain a crystal violet stain under mordanting and subsequent decolorizing conditions (this is the Gram-Stain). The major basis for this difference is that the wall of Gram-positive bacteria is thicker than that of Gram-negative cells so it is harder to extract the mordanted dye. Much of the basic biology of Gram-positive cells is a consequence of the thick wall that has led to different morphology and to different life style effects.

Today we have a glimmer of the strategies employed to form rod-shaped and coccoidal Gram-positive cells. These encompass the sidewalls, the formation of the septum, pole formation, and the inertness of the pole wall once formed. Reasonable ideas for how coccoid Gram-positive cells generally grow have been presented. Additionally, a possible mechanism for how cells can grow as tetrads, as sarcina groups, streptococcal, and as staphylococcal groups have been suggested.

The first breakthrough to an understanding of the "why-and-how" of rod-shaped growth came from studies of *S. pyogenes* and *B. subtilis* (both Gram-positive organisms) (see Koch, 2002). These early works showed both that the pole wall once formed is inert and does not continue to grow or turn over, and that the sidewall of the rod-shaped bacteria grows in an inside-to-outside fashion. This means that the sidewalls of *B. subtilis* grow by adding new layers on the inside and autolyzing (dissolving) the wall's outermost oldest layers. The strategy that results from this pattern of autolysis explained how the Gram-positive rod could grow by cylinder extension as a bacillus, and also it showed that the twisting of chains of the organism during growth resulted from physical forces of this process. Moreover, this realization has led to some understanding of septal formation and conversion of septa to metabolically inert poles.

The first observation was that replication of streptococcal wall was semi-conservative. It was found in 1962 (Cole and Hahn, 1962) that old wall was retained and new wall was laid down centrally in bands of new wall formed between regions of the older wall in streptococci. This resulted in linear chains of cells in which adjacent pairs of poles were of the same age. Consequently, it is the general supposition that Gram-positive cocci generally do not turn over poles once formed but start to insert new wall in the widest diameter of the cells. The staphylococcal "grape" formations in three dimensions are only different in that the division plane successively rotates through the three Cartesian dimensions. Moreover, the formation of Gram-positive cocci is, in principle, understood as resulting from linkages of origin regions of the DNA chromosomes to binding sites present on the polar wall of the cells (Koch and Doyle, 1999).

BIOPHYSICS OF WALL DEVELOPMENT

The turgor pressure inside a bacterial cell creates stresses on the enclosing murein sacculus. The precise distribution of the stresses depends on cell shape as well as on the turgor pressure. Man-made pressure vessels are fabricated and then subjected to pressure, but this is much different than the way it must happen in growing bacteria. The Gram-positive rod-shaped cell, like other bacteria, builds the wall structure *while* the cell is under pressure. It does so in a four-stage process: (i) the constructive formation and insertion of wall material through the cytoplasmic membrane; (ii) its polymerization into new layers of wall adjacent to the cytoplasmic membrane; (iii) subsequently, the resultant layer of wall stretches and becomes stress-bearing causing its conformation to change; and (iv) destructive hydrolysis of external surface wall occurs which permits cell elongation. This last stage is caused by the Gram-positive cell autolysins on the periphery of the wall (Koch, 1990). This means that a layer of the wall polymer laid down at different times in the past is in a different state of strain and intactness. In the following sections the interplay of these physical and biochemical ideas leading to elongation and division and in growth with helical cracks and twisting of the cells will be presented. The wall growth process is significantly different in Gram-negative bacteria where steps (i) (iii), and (iv) function similarly, but an entirely different item; i.e., step (ii) applies (see Chapter 13).

THE GRAM-POSITIVE STRATEGY FOR THE SIDEWALL OF ROD-SHAPED CELLS

The sidewall is laid down all over the cylindrical part of the wall. As laid down and polymerized it is a loose (unstretched, but intact) monolayered fabric. *A priori*, the new layer has to be one layer thick for two reasons: (i) because the forming enzymes are linked to the cytoplasmic membrane and (ii) the penta-muropeptide units emerge on the outside of the cytoplasmic membrane as the result of extrusion through the membrane. However, once a new layer starts to form under an earlier layer, no more penta-muropeptide can be added to the older layer, for simple geometric reasons. As the cell grows longer the fabric becomes stressed and the tension in it increases. This results in the tension being in the Gram-positive wall just outside the cytoplasmic membrane and increasing with distance outwardly from there. This increase in tension does not continue indefinitely. Eventually, the tension becomes too large. Then autolysis ruptures the outer material and the tension falls locally there (Koch, 1990a, 1990c). Because of the distribution of stresses, this results in helical grooves (Koch, 1990b). This allows cylinder wall growth as shown in Figure 12.1.

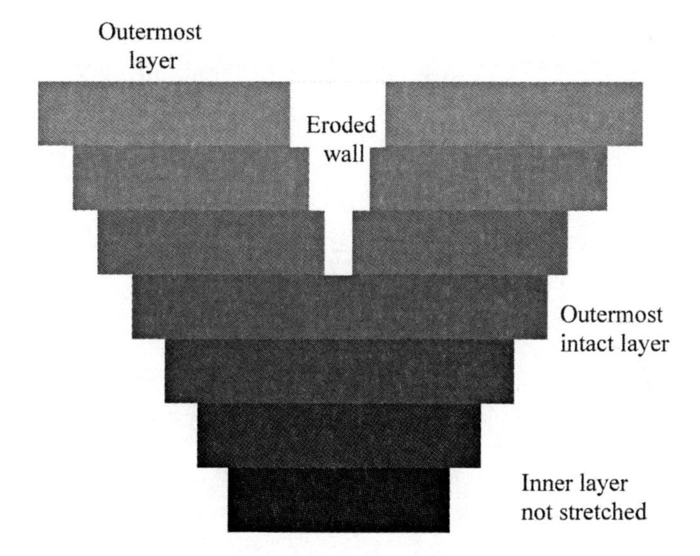

Figure 12.1 Grooves produced by the inside-to-outside growth of Gram-positive rods. As new unstressed layers become added to the murein wall and as they become stressed, they elongate and move outward and form a stress pattern that is not uniform. Where previous grooves have developed and reduced the stress on the surface, there is more stress present at the bottom of the groove than there is where more cleavage takes place. This is a simple case of an analogue of Griffith's law.

THE GRAM-POSITIVE STRATEGY FOR THE POLES OF ROD-SHAPED CELLS

With the quite similar strategy of the addition of successive layers, the cell forms a planar septum, but forms it inwardly (Figure 12.2, Panel A). As the septum closes, it is also split into two halves starting from the outside inwardly by surface autolysins and each of the halves bulges to become a pole (Figure 12.2, Panel B). This expansion occurs without addition of new murein. Notice that the planes of murein extend in different orientations on the cylinder part than on the poles. This provides a reasonable explanation in terms of the orientation of the components of the murein for the inertness of the poles (Koch, 2002). The integrity and inertness of the formed poles is essential for the growth of Gram-positive rod-shaped organisms like *B. subtilis.*

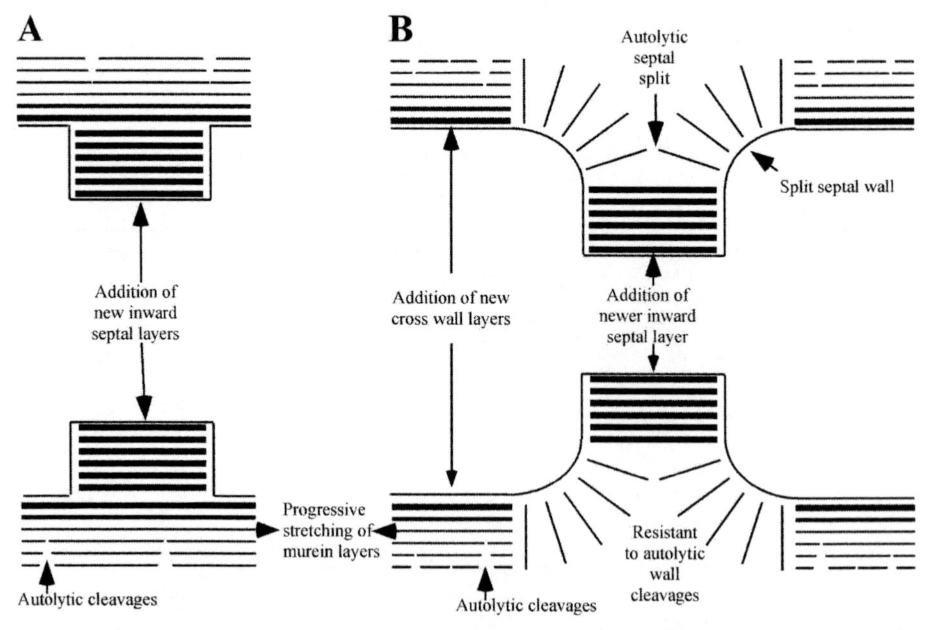

Figure 12.2 Bacillus subtilis septal and pole growth pattern. Panel A shows where new layers are laid down to form a septum. It also shows autolytic cleavages in the outermost wall. Panel B shows the autolytic split that bisects the septum while it is still being completed. Only after the split occurs does the wall stretch to yield a pole in which the layers are perpendicular to the surface and not parallel to it as in the sidewall. This yields, it is postulated, a resistant pole that because of the murein orientation is not subject to the action of autolysins as would occur in the sidewall.

STRESS DISTRIBUTION IN THE THICK WALLS OF GROWING GRAM-POSITIVE ROD-SHAPED BACTERIA

The cylindrical wall of a typical Gram-positive bacterium, while thicker than that of Gram-negative cells, is still thin (25 nm) compared to the cell's radius (400 nm). Mechanical engineering analysis of such a shaped pressure vessel with a uniform wall thickness demonstrates that the circumferential (or hoop stress) is twice as great as that in the axial direction (see Koch, 2001). The hoop stress is greater than the stress at any other angle, and therefore, in an ideal homogeneous wall, cracks would tend to form perpendicular to the long axis of the cell.

However, because the cell is a growing and elongating system, the cracks will not be perpendicular to the long axis. This is because of the accumulation of partially stress-bearing, partially fragmented, older wall outside the most stressed intact layer. Wall grooves that are located in the wall surface tend to swerve and form a helical pattern. The physical process is outlined in Koch (2002). Basically the helical cracks in the outer wall lower the wall strength and favor the addition of additional cleavages at the base of the grooves. This maintains a system of grooves as the cell grows and elongates. It means that once a system of helical grooves arises by chance, it is propagated though further growth. However, if something very dramatic happens, then it is possible that the opposite pitch might arise and be subsequently propagated.

The pitch of the helical grooves is increased by the elongation stresses in the outer layers and is decreased by the hoop stresses in the inner layers. The balance between these processes allows coherent helices to form which elongate coordinately with cellular elongation. No new cracks will form and the existing cracks, given their dynamic character, will elongate with constant pitch. Otherwise either the spacing between cracks will increase and new cracks will appear, or old cracks will make tight spirals and finally stop while new cracks open up in between such sites.

In this model, presented in Figure 12.3, the three key facts about the physiology of Gram-positive rods fit together with considerations of stress distribution and of autolysin due to stress on their peptidoglycan substrate during growth activation to generate a system of helical cracks in the outside layers of side wall.

In the next section it is shown that this growth process leads to rotation of one end relative to the other. Another aspect of this physiology of the Gram-positive rod is shown, this is that *B. subtilis* grows with twisting that is actually caused by the twisting of the cylinders of cells as the turnover process enlarges surface grooves enlarge.

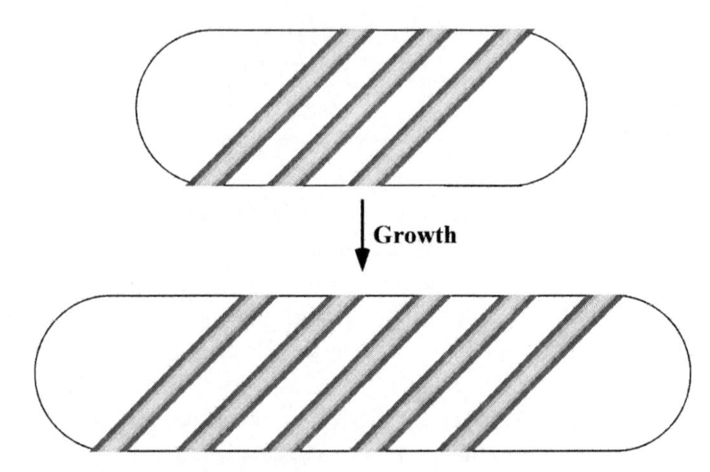

Figure 12.3 Growth with propagation of helices. This figure diagrams the elongation of a helical grooved cell. This leads to growth of the helical grooves of a constant angle with respect to the cell body.

AUTOLYSINS AND TURNOVER

Gram-positive rod-shaped cells have a special feature; this is the activity of their autolysins. In *B. subtilis* there are two major ones, an amidase and a glycosaminidase. The latter is an enzyme like lysozyme, but one that cleaves the saccharide chain between the NAG and the NAM instead of the other way around. The amidase cleaves the muropeptide chain between the D-lactyl group of the NAM and the L-alanine. The evident reason for these hydrolytic enzymes to be present in such large amounts is because of the way the rod grows; i.e., by the inside-to-outside mechanism with eventual destruction of the murein in the outermost layers of the cylinder portion. This was explained above.

The growth of rod-shaped cells is in contrast with the way that Gram-positive cocci form their almost spherical cells. The latter do not turn over thick existing polar walls, instead they form only a septum and split it. On the other hand, Gram-negative cells form only a thin murein wall, and the processes of turning the wall over is different and a very special process.

The name "autolysin" was coined because Gram-positive cells can destroy themselves; this happens particularly during the stationary phase. This is, at least in part, because the autolysins that are necessary for growth can also destroy the cell. This destruction may be a fringe benefit in that DNA is liberated and surviving *B. subtilis* cells can take the DNA up and become transformed. This may allow the cell lines to evolve more rapidly.

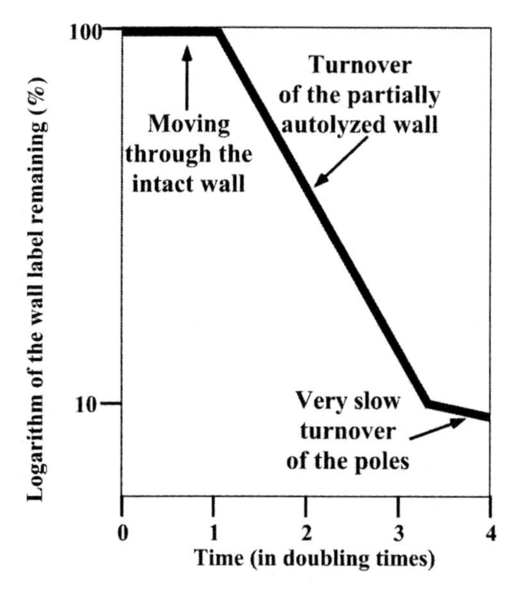

Figure 12.4 Turnover of wall label in a Gram-positive rod-shaped organism. Because of the inside-to-outside manner of cylinder growth, the incorporated label is retained for a generation and then lost exponentially. Finally, only the label in the poles remains.

The inside-to-outside strategy for elongation of the Gram-positive rod leads to an interesting and unique pattern of turnover of the wall. This unusual process has been established by special techniques to stain the cells, by electron microscope studies of movements of a heavy atom-label through the thickness of the wall, and by tracer turnover studies with ^{14}C. In the latter case, many experiments have been carried out in which a precursor of the wall was administered as a pulse and the radioactivity of the cells followed with time of chase. In most metabolic turnover studies, say with a radioactive amino acid, the initially taken up radioactivity is progressively turned over and the radioactive protein taken up is lost from cells in an exponential way. But in the Gram-positive case with a murein label (Figure 12.4), the amount of radioactivity does not decrease significantly for a whole bacterial doubling time. It is then lost exponentially with a half-time equal to the doubling time of the bacteria. After the radioactivity decreases to about 10% of the initial value, this process no longer occurs, and the radioactivity stays nearly constant. Thus, the rate of loss abruptly decreases to almost zero. The interpretation of this pattern is clear. Initially the radioactivity is incorporated into the wall layer adjacent to the cytoplasmic membrane. New layers are deposited underneath and old layers removed from the periphery and consequently, initially, no label is lost from the cell. The labeled layer moves through the thickness of the sidewall as the cell grows. However, only when the labeled layer reaches near the outside does some of the radioactive material become liberated. Because of the grooves the loss is not discontinuous, but exponential. Finally the only radioactivity remaining is in the poles and, at most, only very slowly lost.

A MECHANISM FOR A ROD-SHAPED CELL TO FIND ITS MIDDLE AND SEPARATE SEGREGATED DAUGHTER CHROMOSOMES AND CELLS

A model for cell division is presented in Figure 12.5. It was first published by Koch *et al.* (1981) to account for the precision with which the rod-shaped cell finds its own middle. This allows it to produce nearly identically sized daughters. It depends on the "theta" mechanism of chromosome replication. Replication of chromosomal DNA starts from the origin and proceeds by bidirectional replication until the terminus DNA region is reached. The mechanism presented here accounts for segregating chromosomes and also accounts for the precision of dividing cells. The elements of the centering process are shown in Figure 12.5. At or near the poles of the cell there are special sites, labeled O and O′, that can bind either the origin region, called *oriC*, or the terminus portions of the chromosome, called here *ter*. After a new round of replication starts, a second copy of *oriC* is formed, one copy of *oriC* remains attached (or dissociates and reattaches to the original site), and the sister copy of *oriC* finds the equivalent site at the other end of the cell and displaces the *ter* region DNA previously bound

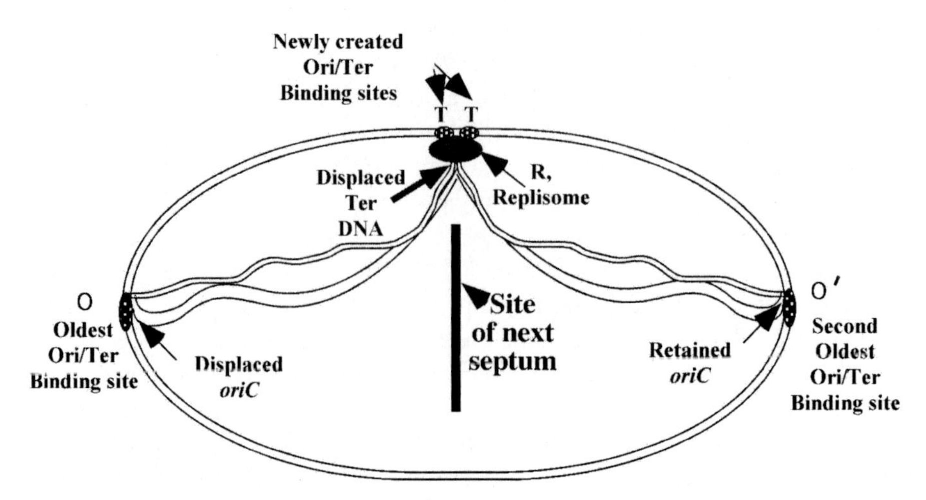

Figure 12.5 Chromosome replication and establishment of next septal location. See text.

there. Then the displaced *ter* DNA, probably with other factors, becomes attached elsewhere to the cytoplasmic membrane, although not in a fixed position, but soon becomes attached to where DNA replication is taking place. R indicates this site as an abbreviation for replisome. Because the replicating chromosome as the DNA as synthesized is in a symmetrical structure like the Greek letter "theta", the proteins associated with it and the secondary structure of DNA creates forces that lead, at least by the time of completion of chromosome replication, to the centering of the replisome and the terminus at the middle of the cell cylinder. In this model, completion of replication triggers septation in the Gram-positive organisms (and the formation of a constriction in Gram-negative organisms). In both types of organisms it leads to the formation of two new membrane-bound binding sites that can act to bind either the origin or terminus. This model is a descendant model of the replicon model of Jacob, Brenner, and Cuzin (1963) but does not require that the growth of the wall take place only in the center of the cell. In fact, elongation of the wall does take place over the entire cylindrical part of the cell in *B. subtilis* (Merad *et al.*, 1989).

Chapter 13

Wall Growth Strategies for Gram-Negative Cells

The Gram-negative cell has a thinner murein wall than the Gram-positive bacterium. It does have an envelope with four components: The cytoplasmic membrane, the monolayered murein wall, the periplasmic space, and the outer membrane. Each of these has many functions to perform. However, only the murein wall is strong enough to give the cell shape and resist turgor pressure. The foremost questions about the wall are: How thick is it? How metabolically stable are the sidewall and the poles? How elastic is it? How stretched is the murein in a growing cell?

Because the wall of the Gram-negative cell is so thin, wall enlargement requires insertion of new murein units (disaccharide penta-muropeptides) into portions of the wall while it is under stress. This requires very special processes that need to be under the critical control of the bacterium. Most importantly, this requires "smart" autolysin action". The job of autolysins is to cleave the murein to allow enlargement, but, of course, it has to be done very carefully. One suggestion to account for this is that the autolysins may be allosteric enzymes and their function has been arranged so that they only act if cleavage will not cause harm. A second model is that the autolysins are bundled with the synthetic part of a complex system for wall enlargement and result in new wall being formed as the old is being cleaved and/or turned over. Therefore very complex holoenzymes have been postulated. These holoenzymes would require a number of proteins to be exported through the cytoplasmic membrane and then aggregated into a multiprotein structure in order to function safely. A third, and recent, model (the Nona-muropeptide Stress Model) depends on the range of conformations achieved by a cross-linked nona-muropeptide under different intensities of tension. The nona-muropeptide on being newly linked into the stress-bearing wall is not stressed, but responds to increasing stress on it and elongates by altering its conformation. This change increases access to the "tail-to-tail" bond of the muropeptide as well as access of new penta-muropeptides to the unbounded groups on the nona-muropeptide. This model could function in a quite simple manner compared to the other two earlier models, mentioned above.

Although the cylindrical sidewall contains both old and new murein, the recent observation (see Chapter 11) is that the sidewall is composed of patches of old and new material. The size of the patches is not constant and ranges roughly to about 100 nm. Consequently, these patches consist of hundreds of oligoglycan chains both across and in length. In toto, this finding is inconsistent with the earlier models in the literature about how the sidewall and poles of the Gram-negative rod grow and divide.

The understanding of the mechanisms that allow the sacculus to grow and divide are important, however, it is the mechanisms that control (or are controlled by) the replication of the chromosome so that the process of cell growth and division and chromosome replication, which are equally important, will occur harmoniously.

THE GRAM-NEGATIVE WALL STRUCTURE

The structure of the Gram-negative wall is shown diagrammatically in Figure 13.1. Although we usually talk of three layers there really are several more. Starting from the inside these are:

(1) The two leaflets of the cytoplasmic membrane. Imbedded in this layer are proteins, some of which bridge the membrane and others that do not. Present are special molecules like the bactoprenol that aids the exporting of the penta-muropeptide from the cytoplasm and proteins for transport processes. There are several different processes, moreover, for protein export to the periplasmic space and beyond.

(2) The peptidoglycan layer. This has to completely surround the cell in a covalent network. It must also contain special proteins for export function to allow certain proteins and toxins to be extruded through the murein.

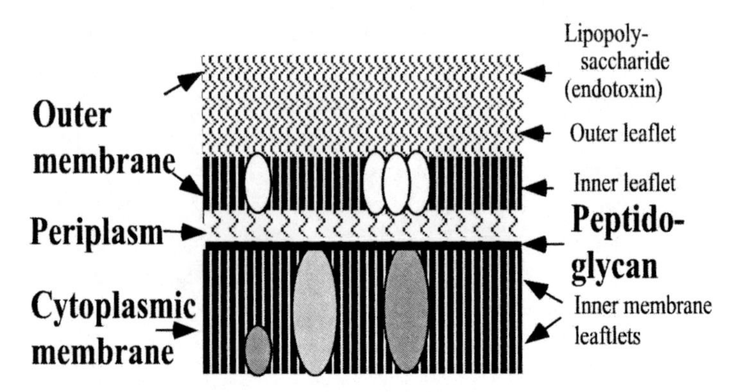

Figure 13.1 The Gram-negative wall in cross-section.

(3) The periplasmic space. This region contains many enzymes needed for digesting molecules for easier uptake and utilization. It also contains molecules to aid in the transport process.

(4) The outer membrane. This membrane also contains molecular assemblies, called porins, to permit small molecules from the medium to enter and leave the periplasm. This membrane is composed of two leaflets with quite different properties. The inner leaflet is formed of hydrophobic phospholipids, while the outer leaflet contains special hydrophilic carbohydrates. Together these layers prevent the entry of a variety of kinds of molecules. In addition there are molecular system aiding in pumping out a variety of kinds of molecules from the cell. The class of molecules excreted by these multiple drug excretion mechanisms contains antibiotics and other fairly non-polar molecules.

THICKNESS, TURNOVER, AND ELASTICITY OF MUREIN WALL

Although there was considerable argument about just how thin the murein wall of *Escherichia coli* is, the experimental data is that the wall thickness is only a little greater than one monomolecular layer. The experiments are not precise, however, since the thickness of a single layer would depend on the tension of the wall and because the experimental methods that have been used have some error. Logical considerations would focus on the idea that the wall synthesis, by insertion of a disaccharide penta-muropeptide substrate into a stress-bearing wall under the action of enzymes (PBPs) that were bound to the cytoplasmic membrane, could only act to extend a monolayer and not to increase its thickness. On the other hand, some muropeptides could be attached at one side of the wall and left dangling temporarily, thus making the wall appear thicker. Also the penta-muropeptides as part of the glycan chains that were pointing above or below the plane of the wall layer would make the saccular layer appear thicker. These could not be structurally effective in Gram-negative organisms and logically would be eliminated by autolysins sooner or later. This, moreover, appears to be the case experimentally. However, before their removal they would contribute to the wall thickness.

THE CONNECTIVITY OF THE SACCULAR WALL

Figure 7.3 shows a stretched, partially stretched, and unstretched view of a nona-muropeptide chain and Figure 13.2 shows a single tessera in a form where its chemistry can be grasped. On the left- and right-hand side, two oligosaccharide strands continue the peptidoglycan; at the top the other half of the tessera would

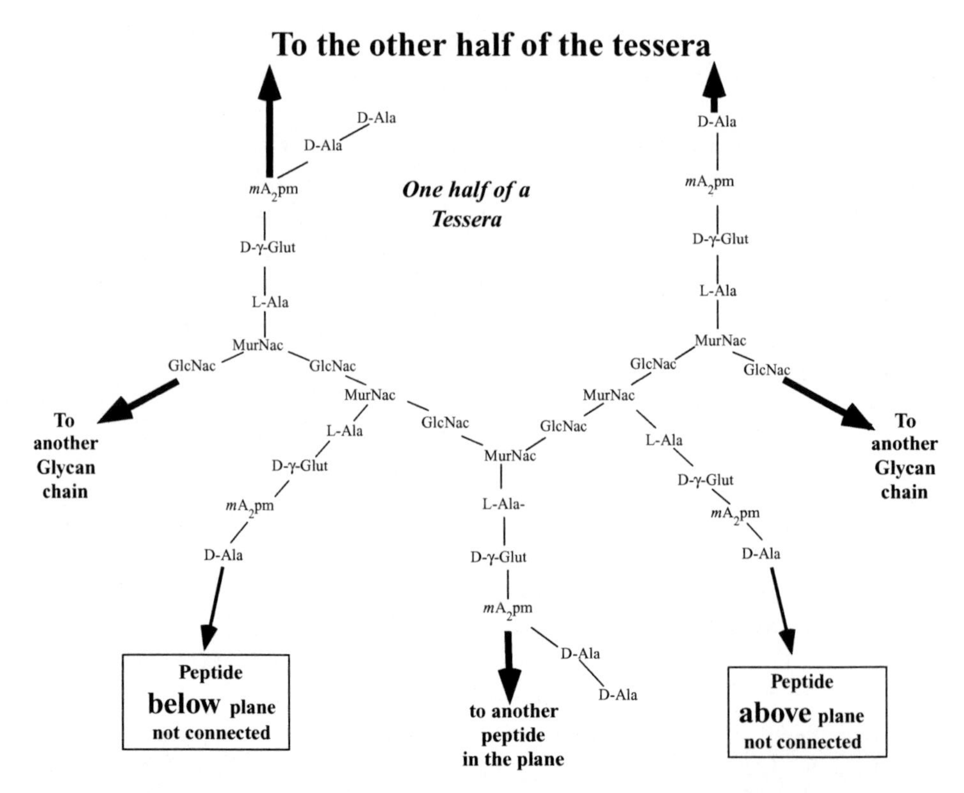

Figure 13.2 One half of a tessera. A tessera in a functional wall is linked to six other tesserae. This takes place via the bold arrows, only half of which are shown. In addition there are muropeptides extending upward and downward, shown by thinner arrows. In a Gram-negative organism with very thin walls, these are not attached to murein and may be degraded from the forms shown here.

be connected, and at the bottom connection to another tessera would occur. Figure 13.3 shows a diagrammatic view of the surface-enclosing portion of the sacculus. It shows a fabric composed of eight tesserae, each of which has been made from five-disaccharides nona-muropeptides. Cross-linked through their peptides at two sites situated eight saccharide units apart to form the unit of wall structure called a tessera. In this figure diaminopimelic acid is abbreviated as mA_2pm.

THE NORMAL STRETCHING OF THE MUREIN FABRIC DURING BALANCED GROWTH

An essential feature of the surface stress theory as applied to Gram-negative (as well as the Gram-positive rods as pointed out in the last chapter)

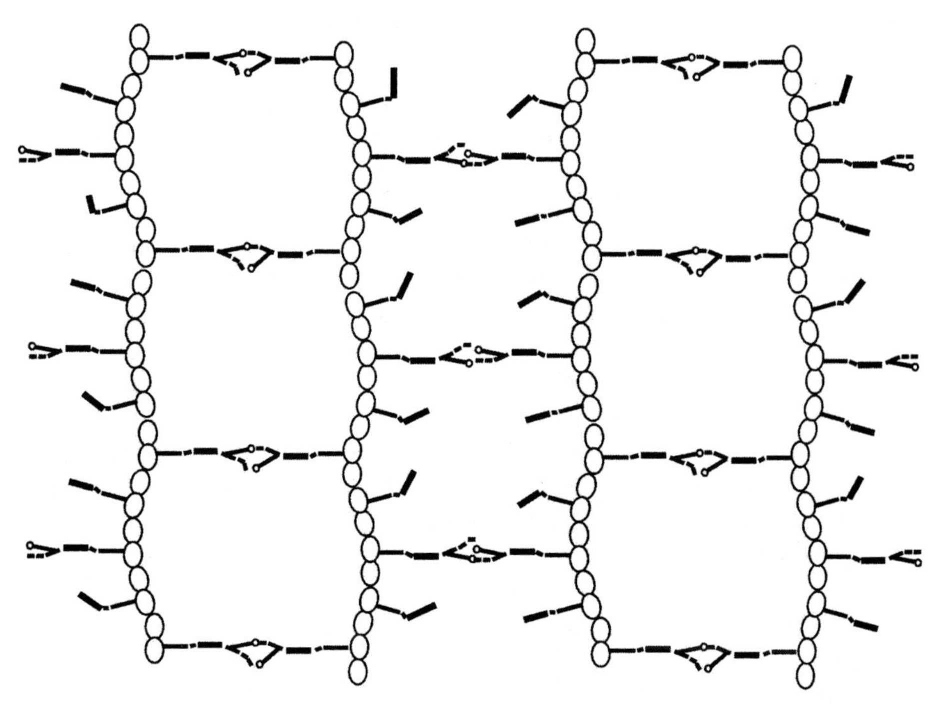

Figure 13.3 A portion of the Gram-negative wall showing eight intact tesserae. Each tessera has two peptides, which are directed above the plane of the murein plane and two muropeptides pointing below the plane. All muro-peptides have been shown (arbitrarily) as having had the D-Ala-D-Ala peptide removed.

is that a newly inserted element of peptidoglycan wall that has been linked is not under stress. However, when subject to stress by subsequent growth it should stretch (mainly, by the bond angles being altered). While expanding is expected, the degree of stretching is not predicted by theory, although some ideas come from experimental and computer studies of the wall components. A light-scattering experiment suggests that a factor of three- to four-fold in area is possible (Koch and Woeste, 1992), if not impeded by adhesion to the inner or outer membranes. In this section we are concerned with the state of expansion of wall fabric in the growing cell beyond its relaxed configuration. The analogy for this stretching could be made to a stocking stored in a drawer versus one stretched on a smaller or on a bigger foot. Several methods have been used to measure the *in vivo* expansion (see below). They give a much smaller degree of stretching than that obtained by the *in vitro* experiments measuring preparations of sacculi. But these experiments show the unit structure network, the tessera, both in a compact and expanded version.

ESTIMATION OF THE DEGREE OF MUREIN STRESS BY STOP-FLOW TURBIDITY MEASUREMENTS

Many studies have reported the measurement of the change in optical density or some other measure of cell size when the osmotic pressure of the medium in which bacteria are suspended is shifted. Most of these can be faulted on the grounds that the bacteria are able to physiologically adapt to an altered osmotic environment. One study cannot be so faulted. In this case the measurements were made with sufficient rapidity to allow the initial events to be observed, uncomplicated by the compensation due to cellular mechanisms. In this study (Koch, 1984b) a stop-flow device was used in conjunction with a narrow beam spectrophotometer. It was found that pentoses, such as ribose, xylose and arabinose, were most useful for such studies because they would not penetrate the cytoplasmic membrane but had a low enough molecular weight that they changed the index of refraction to only a small degree compared with the non-utilizable hexoses or disaccharides.

On the assumptions that the normal turgor pressure was five atmospheres, and that when the external osmotic pressure was raised to match the cytoplasmic osmotic pressure the cell would shrink but would not plasmolyze, and on the further assumption that the wall was isotropic, one could calculate the stretching in area. It was found that the mean area decreased to 80% of its former value with a shift of five atmospheres. This finding corresponded to the wall of the cells initially being stretched by 11.8%, which is the numerical value of $1/(\sqrt{0.8})$.

ESTIMATION OF THE DEGREE OF MUREIN STRESSING FROM MEASUREMENTS OF GROWING FILAMENTS SHRINKAGE AS THE CELLULAR MEMBRANES ARE RUPTURED

A fully independent approach to the above, which directly measured the degree of stretch of the peptidoglycan fabric, was developed based on measuring the length of growing cell filaments (Koch *et al.*, 1987). The adopted protocol used a temperature sensitive strain of *E. coli* which when grown at the permissive temperature divides normally. However, at 42 °C it grows as a filament so that it becomes long enough such that its length could be accurately measured in the phase microscope. After three to four doublings the cells were introduced into a rectangular glass capillary that had been previously treated with a polylysine-containing solution to bind the bacteria to the glass surface.

For the experiment, the fluid in the capillary was changed to one that would disrupt the inner and outer membranes. For our purpose, solutions of sodium dodecyl sulfate (SDS) in the growth medium were used to destroy the integrity of the cell membranes. The average shrinkage measured was 17%. If the circumference shrank by the same length, it was concluded that the area of the sacculus of the growing cell was 45% greater than in the relaxed state.

SIGNIFICANCE OF THE *IN VIVO* EXPANSION OF THE SACCULUS ABOVE THE RELAXED STATE

Since both approaches yield much smaller degrees of expansion (20% and 45%) above the presumed relaxed state under growing *in vivo* conditions of 300% that are observed by the physical and chemical manipulation of the purified sacculi preparations *in vitro*, there is a quandary. The possible errors in the two *in vivo* methods are quite different. The contraction of filaments might be underestimated because of adhesion to the microscope slide, but that is not a problem with the stopped-flow technique. The stopped flow involves use of physical theory and application of corrections for the medium contribution to the index of refraction. The correction must be applied due to the increase in the osmotic pressure of the suspending medium because of the addition of osmolytes. Consequently an error could be introduced during this process. The filament contraction method is not affected by distinctions between the periplasm and the cytoplasm, but there is some ambiguity in the light-scattering measuring procedure. (Light scattering measures the biomass contribution above the water content, and even if the cytoplasmic membrane plasmolyzed and the outer membrane did not alter in shape or size, there would be little change in the light-scattering signal unless substances leaked out of the cell.)

Now, I would amend this statement to apply only to the “smoothed” surface area, because we now know that when the cell shrinks the lipid bilayer is not capable of shrinking, so the wall envelope must wrinkle. Therefore with either the osmotic challenge or the detergent treatment, the murein layer may not be able to achieve its equilibrium conformation. We can, however, conclude that with some small uncertainty the expansion is in the range of 20 to 45%.

GENERATION OF THE ROD-SHAPE AND POLE DEVELOPMENT

From this point in the chapter, theoretical models that concern different aspects of growth will be presented. Of course our understanding is still far from complete, but the outlines of the Gram-negative growth process are emerging.

Both elongation and cell division of thin-walled bacteria must involve adding new wall to the old in such a way that only precisely regulated and targeted cleavages are allowed. A list of important questions concerning safe growth can be posed: (1) What are the physical mechanisms that make both the elongation and constriction reliable? (2) What molecular mechanisms are implemented for stable growth? (3) What details of enzymatic reactions and choice of reaction pathways make growth stable? (4) What forces maintain the cylinder shape? (5) What circumstance leads to the invagination required for cell division? At present there are three relevant models: the Variable-T model (Koch *et al.*, 1982, 1983; Koch and Burdett, 1984), the Three-for-one model of Höltje (1993), and the new Nona-muropeptide stress model for both elongation and division. The three models are not necessarily alternatives, but may represent different aspects of the entire process.

THE PROBLEMS CAUSED BY BEING THIN-WALLED

Growth and maintenance of a rod-shape for a thin-walled single layered organism presents apparently insoluble paradoxes that are indeed confusing. We have: (1) rejected the idea that mechano-proteins and enzymes act to produce a cylindrical shape via a cytoskeleton; (2) rejected templating mechanisms; (3) rejected the inside-to-outside mechanism that functions in the Gram-positive rods; and (4) rejected several more *a priori* possibilities. There seem to be no options left. Worse yet, if only physical forces function, the fact that hoop stress is greater than axial stress would seem inevitably to lead to rounding up into a spherical shape. On the other hand, it is undeniable that the Gram-negative organism elongates over its entire cylindrical region and is so thin that wall units must be inserted into the stress-bearing wall and then quickly come to bear stress. Based on the observation that the peptidoglycan layer is thin, the idea of growth by insertion emerged (Weidel and Pelzer 1964; Verwer *et al.*, 1979; Koch *et al.*, 1981b; Labischinksi *et al.*, 1983). With the development of the surface stress theory and appreciation of the make-before-break strategy for safe growth, it was necessary to reverse the break-before-make order postulated by Weidel and Pelzer (1964).

THE CONTRAST WITH THE GRAM-POSITIVE STRATEGY

It was shown above that the physics of elastic solids would not provide a workable mechanism, due to inevitable bulging, while those of plastic fluid membranes would work but only under proper constraints. To see the problem clearly, consider again the Gram-positive rod solution to the problem. A reasonable mechanism to implement this set of constraints was presented in Chapter 9. A layer of new wall is added uniformly over the inside of the extent wall, which is also very near the outside face of the cytoplasmic membrane. Little orientation for the polymerization can be directed by the stresses in the solid wall immediately above it. Then as the wall elongates its ability to expand has been expended in the axial direction and is no longer available for hoop direction expansion when the layer of wall moves farther outside to where it is no longer supported by more peripheral wall. This pre-emption concept is an extension of earlier ideas of Previc (1970). At this point, the wall is partially supported by helical bands of still-older wall, which gives it a mechanical strength and balances out the two-fold greater stress in the hoop direction that would be felt by a cylindrical shell pressure vessel of uniform material and thickness. (Mechanical engineers frequently reinforce hoses by winding strong wires or straps in helices around the hose. Usually two supporting wires winding in different directions are used. The pitch of the helix is determined by the fact that the hoop stress is twice the axial stress; it then turns out that the optimal angle is $\tan^{-1}\sqrt{2} = 54.7^{\circ}$.)

Chapter 14

Commas, Vibrios, Spirilla, and Helicobacters; Tapered and Branched Bacteria

A beginning of an understanding of coccal and rod growth and of their shape maintenance has arisen with the development of the surface stress theory. The generation of other shaped bacteria is almost completely obscure, two classes, the curved and irregular shaped, will be considered in this chapter. Molecular biology is making important contributions to the study of cellular shapes. Important advances have come and are enhanced by our ability to label the murein of the walls as a function of its age.

Although curved and spiral shaped cells are considered here, spirochetes and mollicutes will be dealt with in the next chapter. In this chapter the bacteria of the epsilon group of the proteobacteria that swim using the external flagella in a quite similar way to E. coli are discussed. They are not cylindrical but have morphologies varying from slight to tight helical structures. The structure of the murein of Helicobacteria pylori suggests a mechanism that may also explain the range of curvatures of many groups of bacteria.

Some cells have bulges and branches, and a possible explanation is presented. These are based on the finding of inert old murein localized in both the poles of E. coli and in discrete patches in the sidewalls. This has led to new paradigms for branches and bulges that are based on the assumption that the patches of old murein in the poles and in the sidewalls are inert, non-turning over, and not enlarging. Another, less likely possibility is presented and is that patches of new material may not be inserted equally on all sides of the sidewall cylinder. This may lead to an explanation of both bulges and curvature of cells.

THE RANGE OF BACTERIAL SHAPES

Earlier in the science of microbiology a major basis for classification was the bacterial shape and, to a much lesser degree, on how the cell morphology changed during the culture cycle. Some strains are pleomorphic, which means that a pure culture exhibits cells of many shapes and sizes. Most organisms however have a more closely regulated morphology. Of course on the average, under balanced exponential growth, the cells grow two-fold in biomass from birth to division. However, even for a well-behaved organism like *E. coli*, conditions can be such that some organisms grow from one chromosome to two while in the same culture other cells are bigger and grow from two chromosomes to four chromosomes. Some organisms in the population, however, do switch from one mode to the other during growth.

Many bacterial organisms alter aspects of their physiology and morphology to adapt to the environment, most particularly, to adaptation to life inside a host. The host can be an animal, a plant, or a protist. Changes may result from a change in life style from that in a liquid suspension into life in one host, into life in another host, into life on surfaces, or in fact, many other circumstances including obligate life within a eukaryote organism. The range of circumstances can alter the bacterium's physiology, presumably by activating or repressing genetic programs carried by the bacteria and also by *de novo* mutation and selection.

Many bacteria are curved to various degrees. There is a continuum of curvature types. Between the straightest bacterial rods and the tightest of helical structures, there are commas, vibrios, spirilla, and especially members of the epsilon group of proteobacteria (Figure 14.1). Quite different are the spiroplasma,

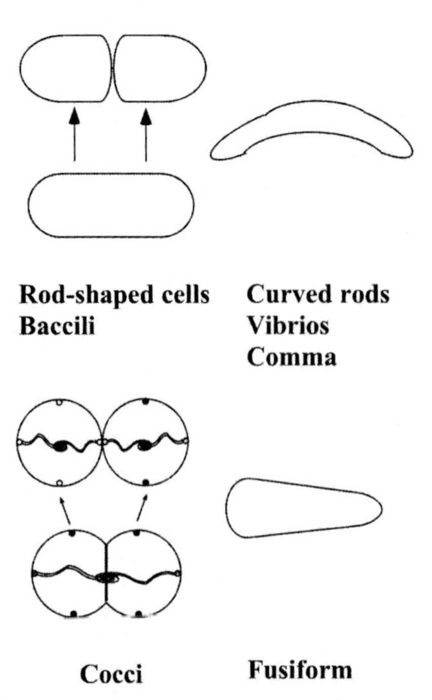

Figure 14.1 The range of shapes of the major types of bacterial cells. This collection idealizes the four types of basic morphology. The rod and the coccus are shown in the division part of their cycles. The curved rods can vary in the degree of their curvature and the fusiform organisms can vary in their taper. Cells with bulges and irregular shape are not shown here.

and spirochetes and these extreme coiled cases will be addressed in the next chapter.

MODERATELY CURVED BACTERIAL RODS

Rigid poles appear to be the basis of cylinder elongation, a point made many times in this book. The existence of two rigid inert poles on the ends of a cell does not imply that the rod-shaped portion of the cell has to be straight. To understand this, the reader can form a cylindrical soap bubble by connecting a soap bubble between two rigid solid supports, and then adjusting the air pressure until the cylindrical shape is achieved (see Koch, 1990a and Figure 11.1). I have demonstrated this in seminars, and described it in publications, and in talks. The demonstration clearly shows that rigid ends are needed to allow cylindrical growth. In addition, this soap bubble type experiment can easily be carried out on a bias. The soap bubble on the right hand was blown with the two poles off-axis from each other. This additionally shows that non-colinearity of the rigid poles does not generate curved cylinders (this is diagrammed in Figure 14.2).

Even if the ridged pole surfaces are not arranged collinearly with each other, the plastic bubble is skewed, but still a circular cylinder and, importantly, a non-curved cylinder. However, it must be emphasized again that while the growing wall of a bacterium should obey the physics of elastic surfaces, the non-growing wall, such as a pole, is a plastic (covalently connected) solid which obeys a quite different physics (stretching according to the rules that follow from the physics of Young's modulus).

POSSIBLE MODELS FOR CURVATURE

One of the possible causes of curvature could be the existence of patches of old murein. From the work of De Pedro *et al.* (1997, 2003) it was shown that such patches occur in a variety of sizes in the sidewall. These insertions are more numerous and larger near the old poles (see Figure 11.3). If this occurs such that there are many insertions of small amounts of murein, as the result of numerous discrete events, then the growth by cylindrical elongation would generate rectilinear rod-shaped bacteria (Koch, 1981). On the contrary, however, if the areas of old patches are quite large and correspondingly there are fewer of them, then frequently this would lead to some bends or kinks. One would expect that such a process would lead to a rod with irregular bends or kinks or irregularly curved-walls, but not to regular spirals or vibrio-shaped cells (see below). A related second possibility comes from the observation that newly formed cell division septa are often wedge shaped. For an example, an *E. coli* cell that has

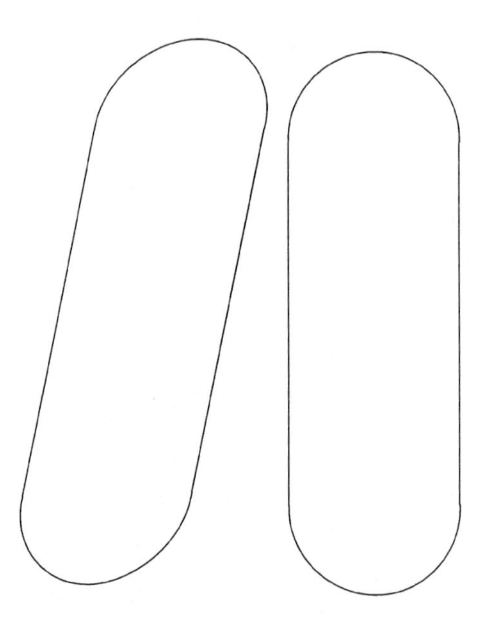

Figure 14.2 Two idealized experiments with a soap bubble pipe. See Figure 11.1 for a photograph of a soap bubble blown between two ridged ends, this process can produce a soap bubble in a cylindrical shape. On the left side of this figure, I show an idealized rod-shaped cylindrical bubble generated as shown in Fig. 11.1. This is formed if the two ends are rigid. It will be cylindrical if the amount of air blown in the bubble is such that the cross sectional area of the bubble is equal to the area of the rigid poles. This construction is feasible, if these rigid plastic ends have been wetted with the soap solution. Then they will not break the bubble when it contacts these solid surfaces. Although not shown in the Fig. 11.1, but one of the rigid ends has to be connected to a tube through which air can be introduced so that a cylindrical bubble can be formed. The amount of air to be introduced would depend on the area of the rigid endplates and the length of the cylinder. The additional issue, relevant to this chapter, is that even if the endplates are not co-linear (as shown on the right-hand side) a right-cylindrical bubble is still produced and is not curved.

been subjected to a one-doubling time chase is shown in Figure 14.3. While this is not a very likely possibility it leads to septal regions that are broader on one side than the other. This yields a pattern similar to that of an *Arthrobacter* cell. These organisms appear to have a jointed morphology like those of a spider's leg. It was from this that the name *Arthrobacter*, was derived. This shape depends on the formation of the septum that starts from a point on the periphery of the cell and spreads bi-directionally around the cell. And thus the septum is broader at its point of origin. In the case of *Arthrobacter* the split of the septum starts at the initiating point and the daughter cells break apart (fully or incompletely) giving a jointed appearance. Non-smooth curvature would result if, instead of splitting the septum immediately, the presumptive, but delayed septum (and maybe some false septa) were just broader at the starting site. It is hard to see how this articulation of division site could give anything other than a disjointed

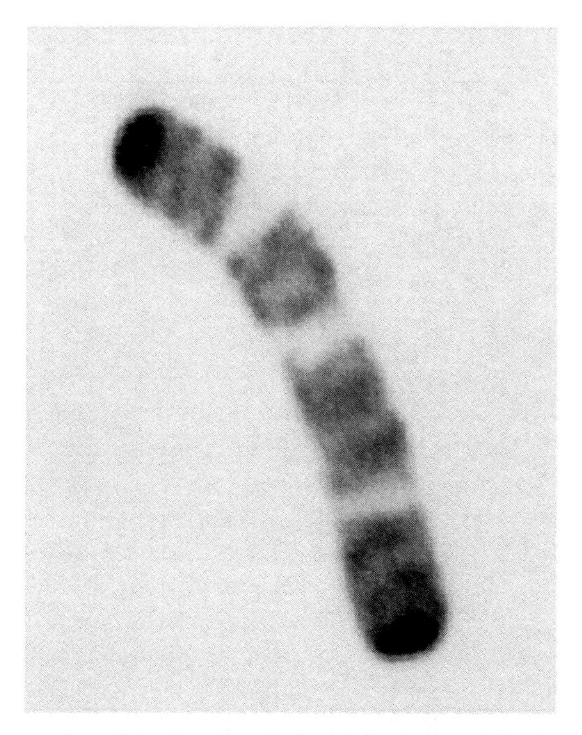

Figure 14.3 The sacculus of an *E. coli* cell chased for one-doubling time that atypically showed curvature. It is taken from the data of De Pedro *et al.* (1997) and analyzed by the methods presented in De Pedro *et al.* (2003). This cell shows evidence of several division sites. The central one is almost in the incipient division stage. It is evident that the curvature results from the sector shape of the forming septal regions. See text. Also note the inert regions at the poles and nearby regions of the rod-shaped cell.

and irregular appearance. Below I suggest a more attractive model based on analytical chemistry of the walls of *Helicobacter pylori*.

COMMAS AND VIBRIOS

Although many other organisms, as for example *Caulobacter crescentus*, have a much more complex shape, we will deal only with the simpler shapes as shown in Figure 14.1.

SPIRILLA

There are members in all the subgroups of proteobacteria, (α–ε), that are spiral shaped. These subgroups of Gram-negative organisms are taxonomically varied with spiral-shaped cells mixed with non-spiral organisms. Because of this large number of different cell types with this morphology we use the term "spirilla", which is not a term used in taxonomy. There are also spiral-shaped organisms in the bacteroides–flavobacterium group. Even though spirilla are quite some evolutionary distances from spirochetes and spiroplasma, they use some of the same tricks. They have internal flagella attached near each end and these must rotate oppositely to generate locomotion.

HELICOBACTERS

The helicobacteria are of special interest here because of the analytical chemistry that has been done on their most famous member *Helicobacter pylori*. It is most well-known because of its involvement with stomach ulcers.

The genus contains a group of bacteria closely related to the campylobacters that are Gram-negative and to various degrees are helical or are curved rods. Its murein is classified as A1γ like the murein of *E. coli* and that of the Gram-positive, *B. subtilis*. *H. pylori*, during growth, varies from a bent rod to a loosely coiled helix of two to three turns or less. During growth other helicobacters have a range of tightness of their helical structures from almost none to very tight (5–7 turns per cell). In the stationary phase, *H. pylori* appearance becomes coccoidal or spherical. The organisms are motile with four to eight sheathed external flagella. There are internal flagella that may support and indeed may have created the helical structure. These have been detected in *H. felis* and in some other helicobacteria. Of the seven officially identified strains that inhabit gastric mucosa, this and one other are unique to the human and the rest are restricted to the stomach of other animals. *H. pylori* cells are human pathogens, have no other known host and mainly inhabit the mucosal layer of the stomach. It is known to cause gastric ulceration and gastric adenocarcinoma.

Because of its transitions in morphology during the culture cycle, this organism's murein composition may be the key to understanding the whole question of bacterial curved rods and helices. In the studies of workers in De Pedro's laboratory (Costa *et al.*, 1999) it was found that short chains (one to three disaccharides long) are present in high amounts. This suggests that they may be links between the longer glycan chains and allow the spacing between the usual longer glycan chains to be greater and more variable than if these chains were linked by only the normal cross-linked nona-muropeptide. These links could allow a more open structure, but be a structure tied together with covalent bonds and not formed of hydrogen bonds. Their presence would allow the murein wall to be flexible and elastic and allow a curved or helical structure to the cell to be formed, instead of only the basic rod-shape structure.

The findings of Costa *et al.* were that the murein of the isolated sacculi contained a large amount (45%) of dimeric anhydro-muropeptides that could be galactosylated. Three things follow from this combination of findings. First, the ability to be galactosylated means that these saccharide groups were at the reducing end of glycan chains within the murein walls. The anhydro group indicated that at the other end of the chain also had to be a chain termination because the formation of an anhydro structure would have blocked further extension of the glycan chain. Thus most of them are only one disaccharide long. The presence of a dimeric peptide indicated that they were cross-linked to other chains as part of the saccular structure of various types. In Figures 14.4 and 14.5 the attachment of such dimeric peptides (nona-muropeptides) are indicated by short bars marked with question marks to indicate that they are glycan chains, but of a unspecified length and function.

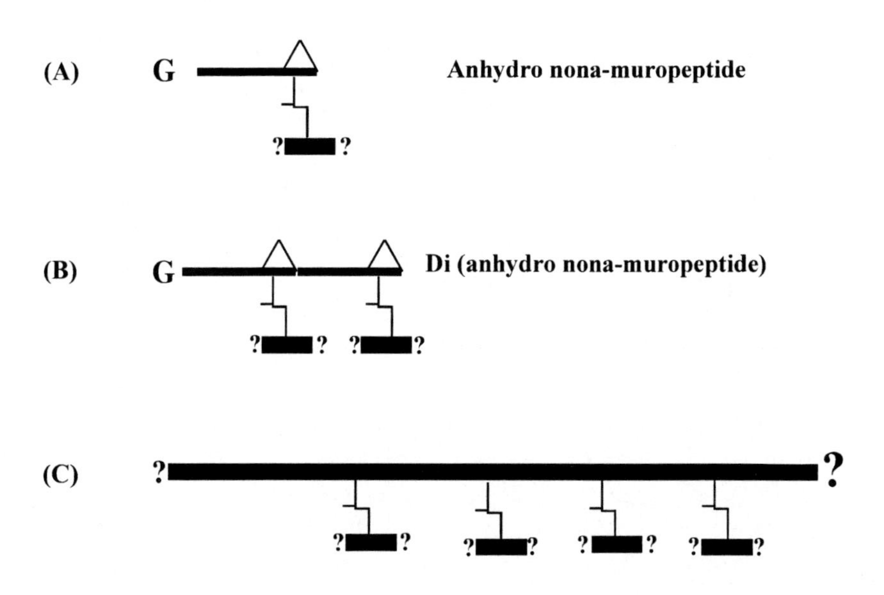

Figure 14.4 Structures presented are typical of the murein of *Helicobacteria pylori*. Figure 14.4A shows a small component that is only a disaccharide long as established chemically (see text), but is connected via a nona- or octa-muropeptide to other glycan chains. These can be of variable lengths. To point out this variability, question marks are placed at their ends. Figure 14.4B corresponds to a structure with two disaccharide linkages. This grouping can function instead of the usual muropeptide to link some longer glycan chains together to form a longer and more flexible chain between them. Figure 14.4C is the normal typical glycan chain with a number of muropeptides that may, for example, be tessera, or may extend above or below the plane of the murein and may be attached to the group as shown in Figure 14.4A and B.

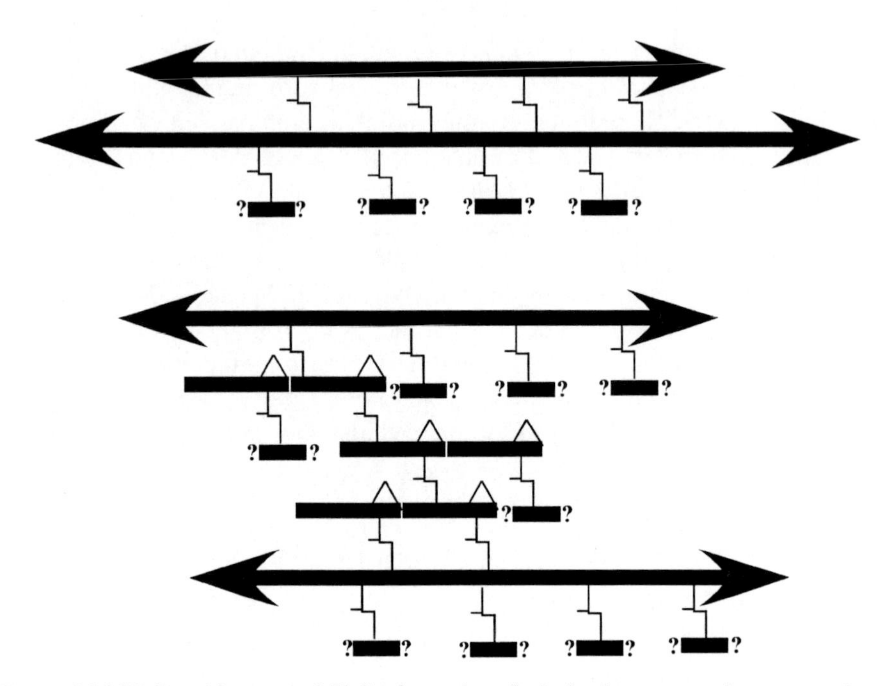

Figure 14.5 Wall architecture of *Helicobacteria pylori*. At the top are shown two glycan chains linked together by octa- or nona-muropeptides. The depiction of the top glycan is inaccurate because each cross-link will point at right angles to that of their immediate neighbors and therefore could not bridge to the next chain. The longer glycan shown connected to it may be bound to other chains or structures like Figure 14.4A or B. The bottom structure shows how structures like 14.4B could function as flexible and longer linkages between glycan such as depicted in Figure 14.4C.

The purpose of the formation and role of the anhydro groups is not fully clear in any kind of bacteria. However, their formation certainly terminates the extension of the glycan chain. It is not excluded that under some conditions the anhydro groups would open up and the anhydro linkage used to allow the previously terminated chain to react with another glycan to cause the chain to be longer. This could occur because the anhydride bond has enough thermodynamic energy to allow the reformation and continuation of the glycan chain.

Costa *et al.* found that 45% of the dimeric anhydro-muropeptides in growing *Helicobacter* could be galactosylated. This implied that these were glycan chains with only one disaccharide, but cross-linked to another glycan chain. Obviously they could not function in formation of a stress resistant sacculus, but had to be peripheral to it. Figure 14.4A shows a disaccharide anhydro glycan with an attached nona-muropeptide. There were 27% of the non-anhydrodimers that could be galactosylated; these would correspond to the disaccharide portion on the left-hand side of Figure 14.4B. These molecules, which are longer than a single disaccharide could function as part of the stress-bearing part of the sacculus. The presence of these short chains, capable of only being linked to the sacculus

Figure 14.6 Shift-up of *E. coli* cells from growth in minimal medium into a rich medium. Figure taken from Woldringh *et al.* (1980).

via their peptides and not by multiple saccharide chains, restricts their usual role in the formation of the murein wall. However, these units that contain at least two disaccharides could play a special role because the multiple nona-muropeptides together could link to other chains of the usual structure. Thus these elements possibly could be the elastic extensions that could permit bends and spirals to be formed and allow spirals with one-handedness to become changed to the other hand. Figure 14.4B shows a chain composed of two disaccharides and bound by the two nona-muropeptides to two other (possibly different) glycan chains. As such it would be the link between chains. Figure 14.4C shows a longer glycan chain linked in a number of places to other chains. Figure 14.5 shows schematically two chains cross-linked not by the usual nona-muropeptides, but by these special links that are longer. These also involve no linkage of one disaccharide with another as in the usual wall fabric. Actually the tension within a sacculus of a growing bacterium is such that with complete murein coverage over the cell the surface would consist of interlocking tessera (Figure 14.6). Mainly they would be short chains like Figure 14.4B. The implication of all these findings is the implication that, although there is still a preponderance of longer chains in *H. pylori*, a large number of very short glycan chains are present.

So how could the short pieces serve for the formation of smoothly curved and helical cells? Figure14.7A shows diagrammatically a section of wall

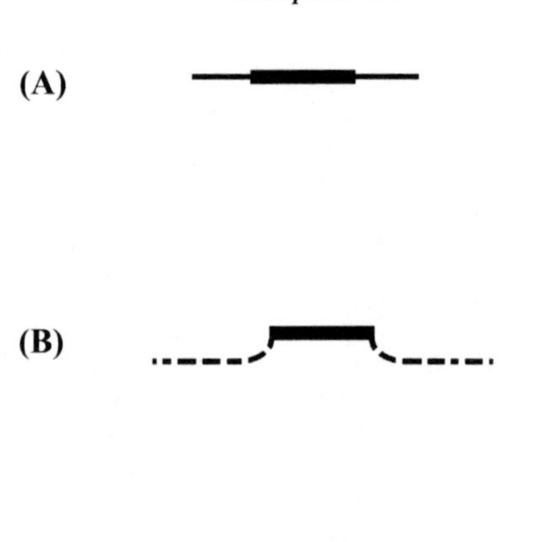

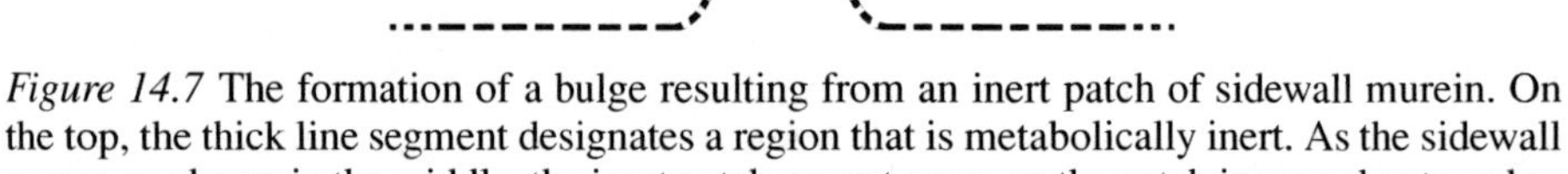

Figure 14.7 The formation of a bulge resulting from an inert patch of sidewall murein. On the top, the thick line segment designates a region that is metabolically inert. As the sidewall grows, as shown in the middle, the inert patch cannot grow, so the patch is moved outward as it is pushed and supported by new growth. This process continues, as shown in the bottom. Other variations on this bulge theme could lead to branches and Y-shaped cells.

composed of tesserae. As depicted the glycan chains are quite long. Figure14.7B shows a section with a range of glycan chain lengths. And Figure14.7C shows a section with the longer linkers resulting from the involvement of the short chains that contain two disaccharides each with a muropeptide attached to some other chain.

TAPERED CELLS

At this point in the chapter, we switch from the consideration of cells with smooth curvatures to other morphological shapes. Some bacteria have a tapered fusiform shape. Although there are no suggestions in the literature to address this problem, several ideas for how this may happen can be proposed. One can be suggested from the experiment of Woldringh *et al.* (1980) with *E. coli*, illustrated in Figure 14.6. These workers found that when a culture grown in a minimal medium was shifted-up to a rich medium, the diameter of the middle section of the cells grew wider than the previous diameter in the first generation. When such a cell divided the old pole was of the original size but the new one was wider. I interpret this (Koch, 2005) as the mechanism that Gram-negative cells have for adjusting their diameter to the quality of the medium. After a few generations, most of the bacteria will become cylindrical, but wider than the parental, minimal-media grown cells. When a culture is shifted-down, then the process that causes the poles to be inert seem to act on only the central part of the pole. For fusiform type growth there needs to be an alternation in the dimensions of the pole's inert region.

To modify the *E. coli* type mechanisms for fusiform organisms, one merely has to add the assumption that not all of a newly formed pole is inert, but a central part is not inert, or is almost inert, so that new sidewall forming from it would grow progressively and become wider, and after division again only the central part of the pole would become inert. In this way fusiform shaped bacteria would continue growing indefinitely.

KNOBS AND BRANCHES

HOW DO BACTERIA BRANCH?

The morphology of rod-shaped bacteria depends on the structure of the covalent linked sacculus. Much of the bacterial shape arises because of the detail geometry in the growth of murein septa that become the poles and its consequences pole inertness to the development of the sidewalls or the new poles of bacilli and cocci. While I emphasize in this book the role of physical force under the aegis of cellular processes, much of current microbiology presumes that enzymes and helical or hoop-like structures control cell shape, though it is unclear (at least to me) how such molecules can determine cell dimensions.

An interesting morphologic shape is the basically rod-shaped cells that have knobs or branches. How they arise is not established at present. However, the surface stress theory and recent experimental and theoretical extensions can be brought together to propose an explanation for this type of morphology. An extension to the biophysical theory of cell structure to be an explanation for the occasional abnormal cell or the special formation of Y-shaped or knobbed cells among cells that are normally rod-shaped can be constructed. It depends on four circumstances:

(i) the conformational changes of the muropeptides under increasing stress, deduced from computer studies;

(ii) the recently proposed mechanism for how murein wall growth occurs, the nona-muropeptide model;

(iii) the timing of synthesis of the sidewall versus polar growth versus cell division site as determined by the De Pedro technique of saccular staining; and

(iv) the roles of penicillin-binding protein 5 (PBP 5) and other D,D-carboxypeptidases as determined in the laboratory of Kevin Young. The resulting suggestion is that the inert patches on the cylinder surface can under some special conditions cause the branching or bulges.

In a recent review, Young (2003) considers possible mechanisms different than the one developed here. The models he considers all depend on a special cellular structure that may be hoops of definitive radii, spirals as scaffolding structures, or an internal cytoskeleton. He does an excellent job of surveying the many ways that cellular morphology can be altered by mutation. The model championed in my publications and in this book depends on the structure of the biologically produced murein and the many ways in which it can be made capable or incapable of furthering peptidoglycan growth.

GENERATION OF *CORYNEBACTERIA*-LIKE MORPHOLOGY IN *E. COLI*

Escherichia coli cells usually grow as rod-shaped organisms with their shape approximating cylinders of constant diameters and of poles that are almost hemispherical. However, this is not always so. Zaritsky and Prichard (1973) were able to obtain branched cells by manipulating the thymine level in a thymine-requiring strain. Åberlund *et al.* (1993) observed branching in minB and intR1 strains. Bi and Lutkenhaus observed peculiar morphology of certain *ftsZ* mutants (1992). Young's group (Nelson *et al.*, 2001, 2002; Young, 2003) has established that an *E. coli* mutant that is specifically deficient in penicillin-binding protein 5 (PBP 5) (but with other deficiencies as well) also grows in this branched fashion. Young's (2003) review cites additional examples of non-rod-shaped behavior of nominally cylindrically shaped cells.

With the technique they developed for labeling murein with D-Cysteine, De Pedro *et al.* (1997, 2003) have demonstrated that the polar murein is laid down earlier than are sidewall and division sites. It was found that in most cells there were patches of old wall in the region in the cylindrical regions beyond the poles. It is proposed throughout this book that the poles and portions of the sidewall are specifically inhibited from normal growth and are key for the formation of normal and, in this chapter, they are the basis of abnormal shapes as presented below.

THE PATCHY NATURE OF THE SIDEWALLS OF *E. COLI*

In addition to the rigid inert poles, there are inert regions in the sidewall of *E. coli*. The observation of metabolic inert regions has led to the concept of patchy sidewalls (De Pedro *et al.*, 2003; Young, 2003), but there is a further aspect of these studies that is relevant here. From Figure 5 of De Pedro *et al.* (1997) and as analyzed further in Figures 1 and 5 of De Pedro *et al.* (2003), it is clear that in the presumptive sidewall region and particularly in regions near the poles there are larger and more patchy areas of old, inert murein. Taking the pole to be approximately hemispherical, one could estimate visually, from image/micrographs of the computer files of the chased old wall that had been fluorescently labeled, that not only was the pole formed of inert, old material, but that a large proportion of cells contained polar-proximal regions that appeared dark and therefore contained old murein even though they were evidently part of the sidewall before the chase was started. This was observed either with one- or two-doubling-time chases before the cells were treated to prepare sacculi.

While these dark sidewall regions may be somewhat diluted with new material, it is clear that the old murein was not turned over or uniformly mixed with new wall. The arguments presented above suggest that this wall, like the pole, has been altered by the D,D-carboxypeptidases. Wall enlargement occurs in other parts of the sidewall. It may be surprising that the cylinder extends as a right cylinder. It is quite clear from Figure 1 of De Pedro *et al.* (2003) that the cylindrical walls are nearly rectilinear even with D-Cys substitution for D-Ala.

THE GROWTH-INHIBITORY ROLE OF POLES IN THE MAINTENANCE OF THE DIAMETER DURING ROD AND COCCAL CELL DEVELOPMENT

The blockage of wall growth in the pole region appears to be a critical aspect for the bacterial way of life and is especially critical for rod-shaped growth. A permanent blockage of wall growth results when a D,D-carboxypeptidase removes the terminal D-Alanine from the free D-Ala-D-Ala of a muropeptide (Young, 2003; Nelson and Young, 2000, 2001). This prevents the muropeptide from interacting with an incoming penta-muropeptide and forming a bond with its incoming diaminopimelic acid residue in tail-to-tail linkage through transpeptidase action. This inertness, according to the surface stress theory, allows the poles to function in cylindrical elongation (Koch, 1981, 2002).

PENICILLIN-BINDING PROTEINS (PBPs) AND CONTROL OF WALL SYNTHESIS

The prime links in wall synthesis are the larger PBPs; these have been assigned numbers PBP 1 through PBP 3. The ones with larger numbers clearly have an important role in wall growth. The small molecular weight PBPs are not wall enlarging but act as D,D-carboxypeptidases. Their main job is to remove the terminal D-Alanine from the muropeptides. Young's group determined that roles of these small PBPs are redundant. Their redundancy could be explained because their complete abolition would block enlargement of murein. Of course, wall growth failure would be fatal. Controlling the (localized) synthesis of pole walls may be so important that redundancy of D,D-carboxypeptidases is needed, since if this action were missing the cell would enlarge as a bigger and bigger entity and eventually self-destruct. Alternatively, the multiple number of D,D-carboxypeptidases enzymes for D-Alanine removal may be needed under different regions or conditions within the wall.

That the critical role of PBP 5 is to prevent branching is interesting, unexpected, and critical. Actually it is the opposite of what might be expected of D,D-carboxypeptidases. Removal of the terminal D-Alanine evidently blocks murein addition, but maybe the process is not always irreversible. I propose the possibility that removal of the terminal D-Alanine is not the only role of this enzyme and that it may also function synthetically to add back D-Alanine. This might happen in several ways. First it may be that PBP 5 is actually a transpeptidase and may have as its two co-substrates both a D-Ala-D-Ala-containing compound available in the periplasm together with a muropeptide in the existing wall from which the terminal D-Alanine had been removed. PBP 5 might also function to bind, but not hydrolyze, thus preventing another D,D carboxypeptidase from acting. Thirdly, the same periplasmic enzyme that functions by exchange to label the old murein in the De Pedro procedure may function to replace the D-Ala, possibly with a glycine and therefore permitting the subsequent formation of nona-muropeptides in the normal way.

THE TRANSPEPTIDASE ACTION OF LARGE AND SMALL PBPs

The transpeptidase reaction used for wall enlargement catalyzes a reaction whose equilibrium constant should be near unity. This is because the bonds, split and reformed, are nearly identical; i.e., a D-Ala-D-Ala is split and residual D-Ala on the peptide is linked to the amino group of the zwitter ion of diaminopimelic acid of another chain. This process takes place without involving a molecule of water. The other reaction product is D-Ala that is bound to the hydrogen atom abstracted from the diaminopimelic acid. The energy difference between these two bonds must be quite small. The reaction is only essentially irreversible because the D-Ala diffuses away from its site of formation. So all that is needed to reverse the reaction is a high concentration of D-Ala free or this residue linked to some peptide together with an enzyme. Seemingly any one of the small PBPs would fill the bill, one with an appropriate specificity could be PBP 5.

THE PHYSICS OF SURFACE EXPANSION WHEN THERE IS AN INERT PORTION OF THE WALL

The surface stress theory was developed in the late 1970s, in response to the realization that a cell does not have a measuring stick to establish its dimensions and does not have the machinery that eukaryotes have to actively move objects within the cell. This theory postulated that a cell could use its previously formed structure, i.e., the poles, as a template to allow the semi-conservative replication of new poles of the same diameter. Given this assumption, the question was how rod-shaped cylindrical bacteria could be formed and reproduced.

While blowing a soap bubble with a pipe is a useful analogy for the growth of bacterial wall, the glass blower's strategy is an even better analogy. Manipulating molten glass with a pipe requires control of air pressure and the temperature that controls the viscosity of the glass. So with a blob of glass at the end of the pipe and slow blowing of air, a tube can be formed with the diameter of the pipe. The process requires skill, but the principle behind it is simple. The bulk of the blob is hot but its mass and viscosity is such that it does not deform with air pressure. On the other hand, the edge of the molten glass where it meets the pipe bulges out. Then because it is thin it cools faster, its viscosity increases as it becomes solid and it behaves as an extension of the pipe. As more air is blown, in the same thing happens again and again. If the pipe is rotated to prevent the temperature from becoming non-uniform in the glass blob, a glass tube can be formed. With skill the tube can be blown to a larger diameter, or it can have bulges and other shapes. But the point is that the wall grows only where conditions permit the glass to expand, and then the physical law permits cylindrical enlargement.

THE INITIATION OF A BRANCH OR KNOB

Consider a restricted area in a surface that cannot grow, but the remainder of the surface can and does enlarge. The consequence is predictable. This can be checked with a soap bubble pipe or by use of glass blowing equipment and is shown diagrammatically in Figure 14.7. The consequence is a bulge.

HYPOTHESIS FOR THE GENERATION OF BRANCHES OR KNOBS

De Pedro's findings make it clear that the pole murein, after being laid down, becomes changed so that it can no longer serve as a site for wall enlargement. If the argument for the role of this process for bacterial growth is correct, then it follows that the same process of inactivating wall growth applies as well to the patches in the sidewall. This ordinarily does not influence the growth of a rod-shaped organism such as *E. coli*, possibly because of the regeneration of a second D-Ala. Such an event could reverse and reactivate the inactive wall. If not, it is possible that a small region may remain inert and new wall would be added all around its periphery without functioning within the inert region. On this basis PBP 5 may be the small molecular weight enzyme that reverses what the other small molecular weight PBPs (including itself) may have done.

THE NONA-MUROPEPTIDE STRESS MODEL AND SIDEWALL GROWTH

Computer studies led to the generation of a new model, the nona-muropeptide model, for wall growth particularly in Gram-negative rods (Koch, 2000a; Koch, 2002; De Pedro *et al.*, 2003). The model postulates that, under the initial low stress conditions, the salt bond, formed of ionically attracted amino (zwitter) and carboxyl groups, would sterically block the approach of the new penta-muropeptides. Two muropeptides would be needed for wall enlargement. The ionic interaction would, additionally, block the approach of the endopeptidase needed to cleave the stressed tail-to-tail bond of a previously formed nona-muropeptide. These events could be initiated under the stretched conformation of the nona-muropeptide resulting from cell growth. Shown in the central part of Figure 7.3 is the conformation that would apply under moderate tension. The high stress conformation is shown on the bottom of the figure in which new muropeptides could enter and bond in tail-to-tail formation with the groups no longer forming a salt bond. According to the new model, the insertion of new penta-muropeptides and cross bridging to the unlinked groups of nona-muropeptide, together with the cleavage of the existing tail-to-tail bond, would together enlarge the sacculus. Consequently, the key point of the nona-muropeptide stress model is the following postulate: When the cross bridge is stretched to a sufficient degree by an increasing cell volume due to cytoplasmic growth, and the two protruding moieties from a recently made nona-muropeptide no longer have an ionic attraction, then and only then can wall enlargement occur. The newly formed cross bridges undergo the same process; therefore the rate of wall growth will be controlled by the rate of increase in cytoplasmic volume. This is a very important feature that any valid model for bacterial wall growth must provide.

With respect to cellular morphology such as bulges, the inert patches on the sidewall can serve as a nidus for their development.

GENERALIZATIONS

Esherichia coli has served as the paradigm for how bulges and branches form. The quarter of a century old surface stress theory provides an explanation with which new experimental evidence fits and meshes very well. The theory is that some parts of the cell are elastic because the murein in these regions is not growing while others are growing and are therefore plastic. The poles of most bacteria serve as the templates for the semi-conservative formation of new poles. Similarly, inert patches in sidewalls serve as the templates for the new poles of branches and bulges. The theory also can account for curvatures and the shape of fusiform cells. The evidence from *E. coli* is impressive and provides the basis of generalizations for many other bacterial and archaeal species.

Chapter 15

Spirochetes and Spiroplasma and the Special Strategies for CWD (Cell Wall Deficient) Cells

Many bacteria grow as curved cells or helices. While basically rod-shaped cells have straight walls, there are others that have a continuum of curvatures from slight bends to tightly curved spirals. In the last chapter, several possibilities for the more gently curved cells were made. In this chapter, we deal mainly with two other groups that have very little or no murein and form tight spirals. These groups are the spirochetes and spiroplasma. Understanding of the cause of these helices comes from the knowledge that they have internal fibrils that rotate or contract/expand to force helical structure and cause movement of the cells. We also deal in this chapter with L-forms of bacteria that are now called cell wall deficient (CWD) cells.

A consideration of the biophysics and physiology of cells with reduced or with only a negligible amount of murein is biologically important and will have increasing importance in the time to come. Consideration of these kinds of organisms may lead to an understanding, not only of how cells with spiral morphology arise but also to an explanation of some human diseases, whose cause had been previously thought to be due to other than microbial infectious agents.

Bacterial cells that have a wall with little or with no murein at all usually live in a habitat that may give them support, such as the inside of a eukaryotic cell. Full appreciation of this necessarily will require an analysis of the strength of a bacterium whose wall fabric is composed of an incomplete set of tesserae.

Many of these CWD cells are known pathogens, and besides the well-known diseases due to them, it has been proposed that some other diseases are caused by CWD cells; such as, multiple sclerosis and ALS (amyotrophic lateral sclerosis, or Lou Gehrigs's disease).

Although we know most about the flagellar motility of *E. coli* (MacNab, 1999), it is quite clear that motility would be quite advantageous to many bacteria. So it is not surprising that a number of quite different motility systems have arisen in different types of bacteria. A group of mechanisms, which generate motility, have been recently presented by Bardy, Ng, and Jarrell (2003). The spirochetes and the mollicutes have totally different systems and although the mollicutes are all related to each other the motility systems have considerable variation.

We will deal with the spirochetes first because their movement mechanism is clearer. Then we will consider the spiroplasma and other mollicutes subsequently. After this we deal with the general case of cells that lack murein wall. If the ideas developed in this section about the physics of bacterial growth and about the evolution and cell wall physiology of these CDW organisms are correct, this could lead to new understanding and new treatments for ALS and a number of other diseases not currently thought to be due to infectious agents.

SPIROCHETES

How spirochetes achieve a spiral shape can be partially explained on a combined architectural, kinetic, and mechanical basis. Available ideas are summarized in this section. The special mechanism depends on internal flagella and on a flexible peptidoglycan wall.

The major genera of spirochetes include *Spirocheata, Treponema, Borrelia*, and *Leptospira*. These genera include the causative agents of syphilis, yaws, Lyme disease, African relapsing fever, periodontal disease, and leptospirosis. All these organisms have a flexible wall and can exist in tight spiral shapes. As they rotate they can move through a viscous medium. These cells grow as coiled filaments because they have rotating axial fibrils. The latter are the equivalent of the flagella found in other species but are located inside the periplasmic space. One (or more) is attached near each pole. The base of the flagella is inserted into the cytoplasmic membrane and the structure enclosing the end of the flagellum is the motor that can rotate it in either direction powered by the protonmotive force. This directionality probably happens as it does in the well-studied case of *E. coli*. For *E. coli*, the power delivering mechanism is anchored in both parts in the inner and outer membranes and the flagellum is external to the cell. For those spirochetes that are mammalian pathogens, the advantage of the location due to flagella inside the outer membrane may be presumed to be, in large part, because of its protection against the immune system of the pathogen's host. This is a viable explanation since flagella in general are quite immunogenic.

However, many spirochetes do not live in organisms with an immune system, so a different explanation for the advantage of this morphology and physiology needs to be made and presented. The usual one is that the spirochete mode of movement gives it the power to penetrate through viscous medium. The fibrils rotate and allow the cell cylinder and outer membrane surrounding the organism to become spiral shaped and to rotate and this causes the cell to move forward. The nature of this kind of motility has become clearer recently due to work from Charon's laboratory (Charon and Goldstein, 2002; Li *et al.,* 2000, 2004). It has been known for some time that the internal flagella are attached near each end of the cell. The flagella at each end rotate in opposite directions when viewed from the point of attachment to the cytoplasmic membrane, but when viewed from the cylinder side of the cell the flagella both ends move in the same direction around the cell and each reinforces the other's action. Because the cell envelope and cell cylinder are flexible, the whole cell undulates. During movement this happens in such a fashion that the whole filament lies in a plane in the case of *Borrelia burgdorferi* but not in the case of some of the other organisms. The cell and filament of *Borrelia* are consequently both planar sinusoids of the same period and direction. However, the plane of the filament rotates in space as the cell goes forward and the planar wave of the sinusoid goes from forward to aft. It is as if the spirochete moves through tissue or soil as a drill bores through a board.

This type of action effectively accomplishes movement and works well in a viscous medium, but it requires that the peptidoglycan murein wall is much looser than in the usual bacterial wall. The presence of some murein must serve an essential purpose because of the well-known fact that penicillin can cure syphilis. Even though the saccular fabric in these spirochetes is sparse, the cell wall has sufficient substance to control the shape of the growing cell and allows it to grow by cylindrical elongation. This implies that the poles are rigid and can support the elongation process as discussed above for other kinds of rod-shaped bacteria even though the sidewall may be flexible. Evidence in part for this is that when these cells change from a rotating to a non-rotating condition, they change from a helical to a cylindrical shape. It is to be noted that, while the cell exists as a planar sinusoidal structure moving through a viscous medium, during growth it divides in the middle and both daughter cells have the same, original, diameter. Two new internal flagella have to be initiated near the newly formed poles.

There are several apparent consequences that follow from this "drill-bit" type mechanism (or strategy). Not only must the new fibrils be initiated and attached during the process of pole formation at cell division (or shortly thereafter), and then the new ones must elongate more rapidly than either of the old fibrils or the cell body itself. In many spirochetes the new flagella must initially grow rapidly and soon overlap those created in an earlier generation growing from the other end. They initially grow in the same way that other flagella grow with additions (or insertions at the base). But these flagella are thicker than those in enteric bacteria because a second layer of protein is added later by another secretion mechanism. The rotational mechanism might be like the well-known protonmotive force-powered rotational mechanism in *E. coli.* Indeed, there are many genes in common between the two.

The control of rotation has to be special since the direction of rotation of the filaments at each end must be of the opposite handedness for translational motion to occur. Consequently, the cell must be capable of rotating fibrils such that one (or one group) goes clockwise and the other flagella or group of flagella at the other end of the cell moves counter-clockwise. Thus, the control of rotation must be quite different than in *E. coli.*

Although thc direction of rotation of the fibrils at the two ends has to be opposite from the point of view of the rotary motors for swimming movement to occur, when they both rotate with the same handedness, the cell flexes instead. This emphasizes the point that the rotary mechanism must be quite accurately and critically controlled so that the sinusoidal shape flagella bend the murein and outer membrane as the flagella rotate. It is presumed that the spirochete switches the mode back and forth to serve the same roles that switch of rotational modes from cw to ccw and back serve for *E. coli* to allow a random change in direction.

Consequently, it is evident that the cell walls for these organisms must be loose structures that can adopt different saccular conformations easily and this must be the molecular consequence of a low degree of cross-linking of murein (i.e., only a few cross-links and/or short oligoglycans). The physiological consequence of low cross-linking is that the osmotic pressure differential with the external medium must be less than that developed in most typical bacteria or rupture would probably occur. As seen from a mathematical analysis, there are alternatives. One is that the cell filament is of small diameter and thus would be more resistant to osmotic rupture. This is actually the case since typically the diameters of free-living spirochetes are in the range of 0.1–0.5 μm. The same point can additionally be made because the wider spirochete cells in this size range inhabit saline or hypersaline environments where the problem is less. If one compares the range of dimensions of spirochete cells with the diameter of *E. coli cells* in minimal medium (0.8 μm), most are narrower. While it has not been measured, it can be presume that the less common, and still wider spirochetes have a lower osmotic pressure. An extreme case is that of *S. plicatilis*, which is 0.75 μm in diameter and forms a tightly curved planar wave. Another exception is *S. cristispera*, which is 0.5–3 μm wide and 30–180 μm long. One explanation could be to presume that it has the lower murein content typical of many other spirochetes, but it may be that its flagella may rotate more rapidly so that a tightly twisted cell may result which might be osmotically stronger, just for this reason. There is one additional way that a cell can persist with weak wall layer and not burst. This is if the cell maintains a high level of an osmotic protectant, such as glycine betaine. However, I know of no reference in the literature to the measurement of the amount and the kind of osmotic protectant compounds in spirochetes.

SPIROPLASMA

The mollicutes are wall-less descendents of the Gram-positive *Clostridia* (Trachtenberg, 1998; Trachtenberg and Gilad, 2001; Gilad, Porat, and Trachtenberg, 2003; Trachtenberg, Gilad, and Geffen, 2003). They are bacteria, but have an "internal cytoskeleton." Two mollicute genera, the *mycoplasma* and *acholeplasma* are small wall-less highly pleomorphic organisms. They are not spiral shaped and they have no murein, and therefore their shape is not rigid. Although they cannot depend on a saccular structure, they do have lipoglycans (with cholesterol in the *mycoplasma*) within the single surrounding membrane; this may give them some rigidity and stability allowing them to maintain a non-round shape. Cytoplasmic cytoskeleton elements emanate from a small region and disperse over the cell body. It is relevant to point out that animal cells also have a requirement for a sterol that gives their cell membrane enhanced resistance to stress.

The third mollicute genus is the *Spiroplasma.* This is the spiral branch of the mollicutes. These cells grow in a corkscrew shape. They too tend to be pleomorphic. They have a pair of flat "cytoskeletal ribbons" within the cytoplasm which are attached to the inside (not the outside) of the cytoplasmic membrane. These seem to be attached over at least most of the length of the cell. These ribbons have been shown by Trachtenberg's group to be involved in the active motility. It is thought that the ribbon contracts and expands causing the cell to be a helical structure of various handedness and lengths. Additionally, this allows the cell to "ram" itself through tissue or soil. The motive power causing the cell to be helical and to contract and expand and to "swim" through a highly viscous environment is at present unexplained.

BACTERIAL L-FORMS AND THOSE WITH CELL DEFICIENT WALLS

Although the bacterial cell wall is an important, indeed, an essential factor for most bacteria, some bacterial species exist, or can exist, in a cell wall deficient (CWD) state. This state includes many bacteria in addition to the two groups that have been discussed above. There are quite a large variety of bacteria that do this as a variant way of life from their normal walled behavior. The L-form is the older term for this and there are CWD forms of many normal bacterial species. The CWD forms may only grow in a structured, viscous, or osmotically enhanced medium. Presumably this is because they need the osmotic and structural support that is afforded by such an environment. Most bacteria while growing under normal growth conditions have an effective saccular cell wall and a fixed morphological shape. Some of these bacteria can give rise to L-forms that have either no murein wall or a very incomplete one. The usual normal forms of different species and the L-forms that can be generated from them, when compared, show the latter to be quite different and atypical and of a quite variable morphology. Additionally, each L-form of a given species can exist in a variety of different but characteristic morphologies. Many of the L-form are pathogens and can survive in this form only in an intra-cellular state within a host cell. In some cases, they are very difficult to eradicate. However, other species can be easily eliminated with suitable antibiotics.

The most peripheral structure of the extreme form of the CWD cell is the plastic cytoplasmic membrane bilayer in a Gram-positive membrane or Gram-negative cell where there is no strong elastic peptidoglycan wall. These phospholipid bilayers, of course, are not as strong as murein and consequently are more easily ruptured or deformed. The implication of reduced strength is clear from the equations for plastic surfaces given above. Therefore CWD cells cannot, in general, be of the width of normal bacterial cells except under very special conditions. Without a murein layer, T, (the surface tension) would be much smaller in the CWD forms, and consequently either the turgor pressure, P, or radius, r, must also be small. Thus either the concentration of cellular solutes is less or the radius must be less. Actually, in a number of different cases both are smaller.

There are some gigantic forms associated with CWD organisms, but these large spherical structures are not cells and their walls do not separate cytoplasm from the external environment. Instead they surround and contain many smaller objects that are actually the CWD cells. There are a number of such specially shaped structures.

Typically the CWD cells are narrower than most bacteria, but they are in the size range of the diameter of most spirochetes. Many examples are shown in the several editions of Mattman's book (1974, 2001) and in most microbiology textbooks and treatises. This narrow shape and small size accounts for the general finding that the CWD forms have higher amounts of phospholipids per cell simply because small cells have a higher surface to volume ratio. This higher concentration is expected from a simple argument: the wall area of a cylindrical shell relative to its volume (if L is the length) is given by $2\square rL/\square r^2L = 2/r$. For example, if the radius of a long CWD cell is $1/5^{th}$ as big as the normal form of the same bacteria, then the phospholipid concentration per unit of cellular mass will be five times bigger.

THE RATIONALE FOR THE EXISTENCE OF CELL WALL DEFICIENT (CWD) ORGANISMS

The compelling reason why bacterial cells form a strong covalent wall was presented above. It might appear that such a saccular covering would be essential for all bacteria in the world today. However this is not so and some bacteria in the modern world as discussed above have no murein wall. Some may have only a weak wall, or a substitute for the murein wall, such as an S layer, where the units are non-covalently associated with each other. These organisms may have cytoplasmic osmotic pressures below the range typical of the walled bacteria, but the measurements of bacterial osmotic pressures are poor at best (see Koch, 2003). Crude estimates of cellular osmotic pressure for normal walled bacteria range from 2 atm to 20 atm.

A murein-poor life style largely constrains the bacterial cells to survive in high osmotic environments and/or in one with a high viscosity. Such an environment is typically found inside eukaryotic cells. Therefore, it is not surprising that some CWD organisms are infectious, and some are important intra-cellular pathogens. It must be mentioned that another alternative for such a cell is to accumulate an osmotic protectant.

THE INSECT CONNECTION

Another characteristic of many permanent CWD organisms is that insects typically transmit them. The classic example is Lyme disease caused by *Borrelia burgdorferi*, a disease that is transmitted by ticks. This suggested a correlation for the evolution and the way of life of CWD organisms and their pathogenesis. I have, therefore, proposed (Koch, 2003b) a connection between CWD organisms and their historical or current existence within an insect vector; this suggestion is made even if the microorganisms may no longer live only in insects.

THE EVOLUTIONARY CONNECTION

Bacteria generally have strong, flexible, elastic sacculi constructed of covalent bonds, and the arrangement of murein of glycan and muropeptides determines the cell shape and prevents cell destruction by the osmotic pressure differential between the cytoplasm and that of their environment. This makes the bacterial cell wall a critical target for destruction of bacteria. It is therefore not surprising that the biological response of other organisms, i.e. plants, animals, some bacteria, and some insects, was to produce potent β-lactamases and lysozymes to block bacterial wall synthesis or to destroy the murein walls once formed. For example, human tears generally have adequate levels of lysozyme to hydrolyze and destroy the wall of most Gram-positive bacteria. A variety of small molecular weight substances, such as the fungal-produced penicillin, also block wall growth and kill cells. But the murein is the target of both types of response and antibiosis by these classes of agents and would be inappropriate in the absence of the sacculus.

Insects have high levels of lysozymes, not only for digestive purposes but also to protect themselves from bacterial infection (Jollés, 1996; Koch, 2003b). However, although these lysozymes allow insects to destroy many of their pathogens, some bacteria have developed a unique counter strategy. This very effective response is simply not to have a murein sacculus at all. Seemingly there would be a large "minus" to this strategy; such organisms would have difficulty surviving in a low osmotic pressure milieu. This, consequently, tends to favor their existence as intra-cellular parasites or their restriction to a relatively high osmotic environment, such as the haemolymph of insects.

INSECT OSMOTIC PRESSURE

An insect's haemolymph has a higher osmotic pressure than is typical of mammalian blood. This fact accounts for the reason that some CWD organisms from insects can only be cultivated in high ionic strength medium and/or high concentrations of proteins, such as found in insect haemolymph. A number of such cases are reported in the third edition of Mattman's book (2001).

If the ideas developed in this section about the physics of bacterial growth and about the evolution and cell wall physiology of these CDW organisms are correct, this could lead to a new understanding and new treatments for ALS and a number of other diseases not currently thought to be due to infectious agents (Koch, 2003).

COULD AMYOTROPHIC LATERAL SCLEROSIS BE CAUSED BY A SPIROCHETE?

It has been suggested that spirochetes may also be responsible for a number of other diseases other than those listed above, including multiple sclerosis and amyotrophic lateral sclerosis (ALS) (Mattman, 2001); however, this has not been conclusively proven. The thought that these mammalian diseases could have a bacterial etiology is no longer novel since it has been shown that helicobacter is the agent that causes stomach ulcers. I have proposed an explanation for the biology (Koch, 2003b) of ALS that builds further on Mattman's ideas. She reported that spirochetes have been cultured from 18 cases of ALS. This point needs experimental clarification. This idea suggests that, on the first detection of symptoms, antibiotics be administered to patients in the earliest stages of the disease.

Chapter 16

Coccal Versus Rod-Shaped Cells, and the First Bacterium

Two very typical bacterial shapes are the coccoid and the rod-shaped bacteria: the spherical and the cylindrical. Their abundance suggests that both are very effective growth forms and that they probably are simpler than other shapes to generate. Moreover, it can be argued that one or the other might have been the first member of the Domain of Bacteria. It is argued that, probably, the first bacterium was a Gram-positive bacillus.

The numbers of different shapes that bacteria do assume is not large. This lack of variety of morphology eliminates an important taxonomic way of identification that is often used by students of any type of eukaryotes and not available here. The coccal shape is evidently the simplest shape and might be presumed to have been the shape of the first bacterium. However, evidence in favor of a bacillus form as the first bacterium has been collected (Koch, 2002). Of course, in order for bacteria of particular shapes to have evolved there had to be the development of the sacculus to support anything other than a spherical default shape.

THE FIRST BACTERIUM RESULTED FROM THE DEVELOPMENT OF THE SACCULUS

The first living cell, sometimes called the cenancestor, arose most likely slightly more than 4 billion years ago, while the Last Universal Ancestor (LUA) probably emerged 3.1 billion years ago. The LUA stage was when the Bacteria separated from the Domain of Archaea–Eukarya and any remaining prokaryotes (which, due to competition, are no longer in existence).

Bacteria became a distinctive domain when its first member developed an adequate cell wall to protect itself against an internally generated osmotic pressure. It is my belief (as discussed above) that Bacteria emerged as a distinct domain when its first member developed a murein sacculus. When life started at a much earlier time, protection against turgor pressure was a non-problem. As time passed however, turgor pressure in cells increased because the concentrations of intracellular constituents increased as cells became metabolically more effective in growing and carrying out more versatile chemical reactions. By forming a sacculus—that is, a relatively rigid, but completely enclosing external wall—to withstand that elevated pressure, bacteria set themselves apart from the remaining organisms extant at that time.

EVOLUTION FROM THE FIRST CELL TO THE LAST UNIVERSAL ANCESTOR

The first cell had to be a minimalist, and had to depend on the environmental availability of those things essential for its growth and reproduction. Even so, three attributes seem absolutely essential for life and evolution: a source of energy and a way to trap it, a capacity for making specific (even if imprecise) functional macromolecules and using them to produce things that the cell must have, and a capacity for replicating (both without error and also with possible changes). Although other properties would be useful and valuable for success and the most important of these gradually evolved, seemingly a strong, stress–resistant wall was not initially required. However, as metabolic capabilities grew, the organismal cytoplasm no doubt became increasingly crowded with an assortment of biochemical ingredients, leading to a progressively greater osmotic pressure. This would make ability for crowded cells to withstand their own turgor pressure absolutely essential.

When a single vesicle incorporated the three mentioned essential qualities, life began. It could replicate on earth because the vesicle/cell could grow and replicate. The primordial living vesicle/cell over generations developed and incorporated a range of improvements. Given enough time (no more than a billion years), a number of different but ever more sophisticated cell processes developed, each of them capable of doing some of the things that most cells alive today can do. The fact that modern cells have so many abilities and chemical attributes in common is the major reason for postulating an ancestor common to all of them. This Last Common Ancestor had incorporated thousands of individual processes and variants of these are in use in cells living today. Despite detailed differences among these properties in different species, the similarities among them are very great, suggesting that they were derived from a cell with a set of common (and possibly identical) skills. This LUA group, or more likely an individual cell, is also called individually or collectively the "Last Universal Common Ancestor" (LUCA) or sometimes the progenote.

Those abilities that developed before the time of the LUA were extensive, essential, and in some cases also sophisticated. They can be categorized into several groups including: (i) processes for trapping energy in usable forms, (ii) transporting moieties in and out of cells, (iii) metabolizing molecules, (iv) synthesizing macromolecules, (v) synthesizing informational molecules, (vi) regulating cell metabolism; i.e., coordinating the cell cycle and balancing the amounts of cell components, (vii) controlling cell shape, and (viii) responding to environmental challenges. Although modern species incorporate many variations, extensions, and redundancies within these eight categories, the LUA surely possessed enough processes in each of those categories to be functional in these important aspects of life.

But what had not been developed and therefore was lacking in the LUA? One critical missing trait was the ability for such cells to withstand much osmotic stress. Of course, also missing was truly effective energy trapping mechanism now present in the biosphere, most notably oxygenic photosynthesis, which clearly arose long after Bacteria originated.

WAS THE WORLD BIOLOGICALLY DIVERSE BEFORE THE LUA AROSE?

Two lines of reasoning suggest that the early world was not biologically diverse and was, instead, monophyletic. The first such line argues that the commonality of cellular processes largely came from the common root of Bacteria, Archaea, and Eukarya, and that differences developed only later. This implies that the LUA, whether it was composed of one or many individuals, was not genetically diverse.

The second such line of reasoning argues that this early lack of diversity reflects the dearth of resources available to these primordial organisms. Throughout the period before the time that the LUA arose, organisms depended on nutrients from abiotic sources, both from chemical reactions, and from extraterrestrial fallout. Only near the end of that period could those organisms begin to use recycled materials from dead organisms as nutrient sources and conversions of one intermediate into another. But a subsequent big change came about when organisms developed other means for harnessing environmental energy, thereby producing or obtaining more, and alternative nutrients, for example from processes such as methanogenesis and the several kinds of photoautotrophy and chemoautotrophy. Some of these new means for harnessing energy also provided metabolically useful carbon resources to build various intermediaries for biosynthesis. However, these energy sources for cell growth only became effective after the time of the subdivision of living forms into domains and biological diversity.

The Russian ecologist G. F. Gause formulated the competitive exclusion principle during the first half of the 20th century, and it became a cornerstone of ecology. According to this principle, the number of species in a habitat depends on the available number of different (unique) resources. Thus, no two species can compete indefinitely while occupying precisely a particular niche within a particular habitat with only a single limiting resource. Consequently, until other nutrients and energy resources were developed and different organisms could specialize on different resources, it was likely that the number of primordial species was limited mainly if not entirely by a restricted form of chemiosmosis for energy. They, therefore, had only one very limiting energy source; this would result in a monoculture. Some scientists argue that primordial organisms might also have developed and derived limited energy from an exotic form of photosynthesis, or energy could be extracted from polyphosphates in rocks. Belief in a monoculture follows because these two do not appear likely and more sophisticated processes such as methanogenesis, aerobic, and anaerobic photosynthesis apparently did not exist before Bacteria and Archaea emerged.

HOW DID DIVERSITY ARISE?

Only two general circumstances would have initiated and generated diversity in a resource–limited biosphere. One is if there had been a novel development that supplied an abundant source of bioenergy. Then energy would no longer be in short supply and limiting and thus permit many different possible limitations for different lines of development. As the biomass of the world expanded, many different species could arise that would specialize in consuming various resources. Methanogenesis, anoxygenic and oxygenic photosynthesis, and many types of chemoaututropy could therefore have broken the monophyleticism. Any one of them could lead to diversity, overcoming earlier limits to growth and thereby meeting this key Gausian environmental requirement in order to generate multiple kinds of stable species. But that appears not to have happened until the time of the LUA at about 3.1 billion years.

The alternative possibility is a special case of the Gausian competitive exclusion principle. I think this is what did happen and occurred when the evolving monophyletic organisms developed a trait that limited their own growth. If more than one response to the osmotic problem resulted this would permit multiple solutions to the problem and many stable species to arise. One possible (even probable) self-created problem of this kind is the results of gradual evolution of a more and more successful organism that had higher and higher cellular osmotic pressure. This problem would have progressively arisen as early organisms gradually became more efficient in terms of their cellular metabolism. This heightened metabolic efficiency necessarily would lead to the risk of the cells rupturing and dying.

Today, a range of organisms from bacteria to plants to animals employ a number of different mechanisms for combating such turgor stresses. However, it is believed that the first such mechanism to develop was what we now recognize as the bacterial sacculus, or exoskeleton. This strong, yet elastic structure covers the bacterial cell and prevents the cell from swelling or rupturing when the osmotic pressure is considerably higher inside than in the outside.

Members of the Domains of the Archaea and Eukarya have developed additional strategies for withstanding high, internal osmotic pressures. These include a pseudomurein exoskeleton for Archaea, structurally similar to murein, but chemically different. The Eukarya and mycoplasma do not have an exoskeleton, however, they are protected by endoskeletons that hold the cell together from the inside.

THE IMPORTANT ADVANTAGES OF HAVING A SACCULUS

Developing a strong and elastic covering outside the cell was not a simple process, nor does it seem to be an economical one (Figure 16.1).

Despite differences among them in different species, bacteria cells with exoskeletons incorporate several general useful properties that include

(i) protecting against too high an osmotic pressure developing in the cytoplasm;
(ii) bacterial cells to have non-spherical shapes;
(iii) transmitting stresses; murein is an elastic substance, like rubber or steel, and when it is not growing it is not a plastic material that can flow;
(iv) timely saccular division;
(v) cell diameters to remain constant; the poles are generally inert once formed and support cylinder extension and septal formation at a fixed diameter for both cocci and rods;
(vi) providing the ability to generate bacterial shapes other than spheres, such as rods, dividing coccus, fusiform, and even spirochetes and spirilla.

Although we now understand a great many details about the genetics and biochemical composition of the murein, the anatomy of the sacculus, and how they it is assembled, we understand very little more. It is likely that the stages of saccular formation are multiple but takes place simultaneously with many components (Figure 16.1).

Critical components of this complex process include

(i) each subunit that is inserted into the growing sacculus must be linkable by three covalent bonds with other such subunits. Two of these bonds are glycoside bonds for incorporating the disaccharide subunits into growing oligoglycan chains. The third is

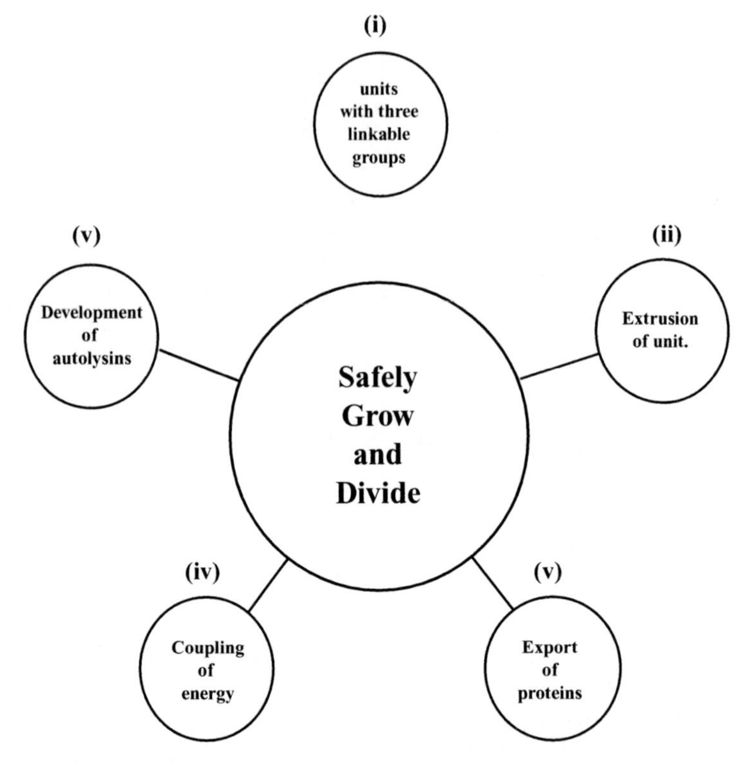

Figure 16.1 Five-component process in murein saccular formation. The minimum set of processes needed to create an exoskeleton or sacculus includes: (i) metabolic transformations to make disaccharide penta-muropeptides by simple modifications of existing enzymatic steps; (ii) extrusion of the disaccharide penta-muropeptide through the lipid membrane with the help of bactoprenol, also known as unadecaprenol; (iii) extrusion of proteins, such as the penicillin binding proteins (PBPs) and autolysins (the transport of these may have been patterned after secretion of exoenzymes needed for saprophytic growth); (iv) exportable energy in the form of disposable pyrophosphate and the peptide bond in D-Ala-D-Ala to support thermodynamic requirements for building and repairing this structure outside the cell proper; and (v) precise control over autolysins to maintain safe growth.

Of course, not all bacteria are the same shape (Figure 16.2). But which was first?

the muropeptide and it can form a tail-to-tail peptide (or amide bond) to cross-link a growing oligoglycan chain to another penta-muropeptide to form the nona-muropeptide crossbridge;

(ii) A. the key step for enlarging thick cell walls of Gram-positive bacteria involves adding layers of new murein adjacent to the cytoplasmic membrane. As new layers are added a layer moves outward for the sidewall growth and inward for septal growth. Autolysis occurs at the external surface of the cross wall permitting elongation. No autolysis (but splitting does occur) occurs on the internal surface including the in-growing septum. A quite different process that is needed for cell division also occurs, however, this causing the splitting of the septum;

B. the key steps for enlarging the thin wall of Gram-negative bacteria involves: (i) converting dimer units consisting of two penta-muropeptides plus another penta-muropeptide into a trimer and then back to a different dimer, all within the periplasmic space; (ii) stretching the nona-muropeptide cross-bridge until it changes its conformation and can accept two new penta-muropeptides (Figure 10.2). This change also exposes the existing tail-to-tail bond that then can be autolyzed. This process depends on the fact that the peptides' cross-bridges can, and do, become elongated as they become stressed.

WERE GRAM-POSITIVE ROD-SHAPED CELLS THE FIRST BACTERIA?

The first bacterium is thought to have been the originator of the formation of a stress-resistant sacculus or exoskeleton covering. This protects the cell from rupturing as a result of stress generated by the turgor pressure that arose from the success of its metabolic abilities. What were this cell's properties? Was it Gram-positive or Gram-negative? Was the cell wall thick or thin? Was the first bacterium a coccus or a rod-shaped bacillus? (See Figure 16.2.)

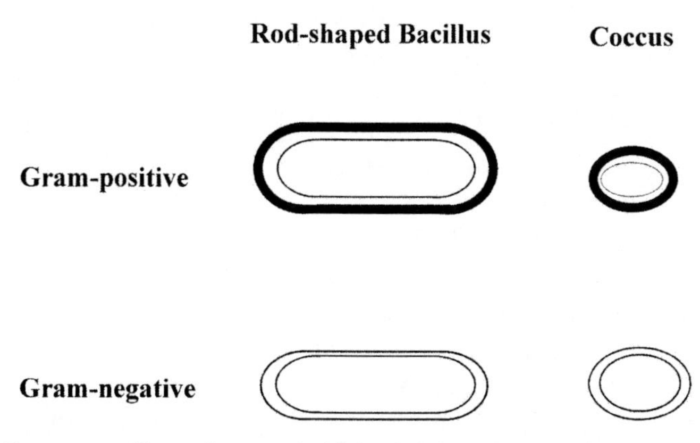

Figure 16.2 Four types of bacteria, one of which might have been the first bacterium. This first bacterium could have had the thick wall of Gram-positive or the thin wall of Gram-negative cells. It could have been rod-shaped or an almost spherical coccus. The text discusses the possibilities of each of the four being the first member of the Bacteria. Note that a variety of more complicated shapes are not included, nor are various kinds of appendages, pili, fimbrae, and flagella, considered at this point.

ARGUMENTS FOR GRAM-POSITIVE ROD-SHAPED CELLS ARISING FIRST

Arguments from three groups, reviewed in (Koch, 2002), suggest that the first cell to separate from the monophyletic prokaryotes predecessors of the bacteria was a Gram-positive, rod-shaped organism:

(i) Seifert and Fox (1998) noted that the rod-shaped structures clustered at the base of the bacterial phylogenetic tree. They compared the morphology of the cells in the various branches of the 16S rRNA tree of Woese (see Olsen (2001)). Seifert and Fox stated; "It [also] seems likely that the last common ancestor of the domain [of] Bacteria was rod-shaped".

(ii) The second group, Tamames *et al.*, (2001) in Vicente's laboratory drew conclusions from the analysis of the 'dcw' (division and cell wall) clusters present in many bacteria. There are 15 genes in the dcw cluster of *E. coli* and about the same number in other species; Vicente and colleagues compared the coding order of the genes (i.e., the sequence of genes on the chromosome) in a variety of species. They found there was a pattern, and found that the cluster was more compact and conserved in bacilli, implying that rod-shaped bacteria came first. These bacilli encompassed both Gram-positive and Gram-negative forms. The bacilli in contrast to the cocci that they studied had a more constant order of the genes on the chromosome.

(iii) A third independent approach is that of Gupta (2001, 2002; Gupta and Griffiths, 2002). He and his group studied the completed, published sequences of many genomes, both bacterial and archaeal, and concluded that Gram-positive bacteria arose first, and Gram-negative bacteria arose from them through a sequence that included several other taxonomic groups.

The method of Gupta (2002) was to look for corresponding regions in available sequenced genomes and find differences that are characteristic of all members of a taxonomic group and that, additionally, are unique to certain, but not other, taxonomic groups. Such differences of "significance" they called "indels" [insertion/deletion]. These indels, consequently, correlate with a taxonomic group. When these persisted in all members of a taxonomic group, Gupta concluded that the grouping had originated from a founder cell generated from another (earlier) phylogenetic group that, by chance, happened to have this particular indel character. It was therefore present in all members of the new group, and then was passed on when a new founder from this old group led to a still newer group. Thus, the character was common to all members of a group no matter what other lines of diversification occurred within a member later on. The important conclusion drawn was that major bacterial taxa arose linearly (in ladder fashion) from each other (Gupta and Griffiths, 2002).

A supporting argument for this order of evolution is that the Gram-negative cells are structurally better able to function as pathogens and to resist antibiotics. This would be in accord with the idea that the Gram-negative cell arose only when there were plants, animals, and fungi to be parasitized and resisted. As a consequence, Gram-negative evolution would have been a later event.

ARGUMENTS FOR GRAM-NEGATIVE CELLS ARISING FIRST

However, the above is not the only opinion. Woese (see Olsen, 2001) argued that Bacteria arose a little bit earlier than the split that led to the separation of Archaea and Eukarya from the existing prokaryotes. The phylogeny of Bacteria that his group derived, which is currently well accepted, does not group the Gram-positives together and does not lump the rod-shaped cells segregated from the cocci.

A major proponent for the idea that Gram-negative cells arose first is Cavalier-Smith. He has presented extensive discussions of the origin of life (Cavalier-Smith, 2001, 2002). He argued first that life started on the outside of a bilayered lipid vesicle that had been produced abiotically. This assumption that life started on the outside avoids the problem of transport across a lipid bilayer of hydrophilic material, but requires that biomolecules of crucial importance remain attached to the outside surface. Later, when life had adequately developed, according to this theory, the vesicle engulfed the living portion. When fusion was completed the first (bacterial) urcell was formed, it was surrounded by two bilayers, and thus the first cell was a Gram-negative organism. The space in between these layers corresponds to the periplasmic space.

WAS THE FIRST BACTERIUM COCCUS OR ROD?

Turning to cell morphology, until a cell develops special mechanisms to establish its shape, the default would appear to be that of a soap bubble. The phospholipid or lipid bilayer-enclosed cell must have had a tendency to be ever increasing in size as the cell grew. Thus unless the cell has devised a cytoskeleton or a strong wall (that is, unless it has an endo- or exoskeleton), it could not have a mechanism to divide to produce new cells, nor cells of a constant birth size, nor cells of any special, but constant shape. If a cell is not attached to any solid object or if has no external constraint, seemingly it must have grown bigger and bigger (until it could no longer grow).

This certainly is not a working strategy for growth and must be wrong. Attachment to surfaces may have helped such a cell to divide, although that would happen irregularly in time and led to irregularly shape daughters. Only one alternative to the above quandary has been suggested for division during the time between the first cell and the LUA (Koch, 1994). This proposed idea is that the lipid constituents of the bilayer were generated in the inside of the cell from molecules that were hydrophilic but then became less polar when they were converted into phospholipid type molecules. As phospholipids or their equivalents they could become inserted into the inner layer of the bilayer and this uneven growth would cause the inner layer to invaginate. Eventually this would lead to division of the cell. This strategy might have worked, but would have been irregular and not have been very effective. It was possibly only a billion years later, when a prokaryote succeeded in completing the evolution of a stretch-resistant fabric, that an effective alternative arose. (Argued here as the creation of the Domain of Bacteria.) Similarly the development of pseudomurein may have led to the creation of the first Archaea. Of course, later the cytoskeleton and contractile proteins arose that together are used by Eukarya.

The growth of bacteria only by the diffuse enlargement of the murein sacculus is insufficient to enable bacterial type growth. It was also necessary to have mechanisms that prevented wall growth in certain regions (the poles), while fostering it in other new poles and sidewalls in an amount needed to accommodate the protoplasmic growth. Such a mechanism would provide a way so that the inert part of the wall; i.e., the poles, could serve as the measuring stick to maintain the width of succeeding generations of cells. For rod-shaped cells it provides support for elongation of cylindrical growth (Koch, Higgins, and Doyle, 1982). Of course, this could occur only if the poles were metabolically inert and rigid.

The implication suggested by this is that the first bacterium, like all modern bacteria (with the exception of mycoplasma) forms two new poles entirely by new synthesis, but subsequently enlarge them no further. It also does not turn them over. This implies that the biochemistry and biophysics must be arranged so that the mature poles are blocked from further metabolism or being turned over (see chapter 10). Experimentally, the size of completed poles is not further modified by insertions nor is it turned over significantly in future generations. It therefore appears likely that cocci only need to be able to form a septum centrally, grow without changing the poles maximum diameter, and then allow, or aid, the septum to split (and to bulge) to form two new daughter cells of the same diameter.

This is the way the coccus, *Enterococcus hirae*, grows (Koch and Higgins, 1984). Evidence from electron microscopy is that at some critical stage of the cell cycle, a septum starts to grow inward (from the site of the previous septa). As it is formed it starts to split from the outside, proceeding inward, and the intervening wall bulges outward, thus forming two new poles. Consequently, the new pole has the same diameter as the old one and in the next generation, these nascent poles become the templates for the still newer nascent poles. This means that the diameter of poles in a culture in balanced growth is remarkably constant and is in fact ±5% (Higgins, personnel communication)).

Obviously, growth of cocci is simpler than what is necessary for a rod-shaped organism that must form the cylindrical walls as a separate and as an additional process from septal formation and splitting. This would suggest that the first bacterium was a coccus. This thought would not be in accord with the logical extension of the ideas of Woese (2000), of Seifert and Fox (1998), of Vicente's group (Tamames *et al.*, 2001), or of Gupta's group (Gupta and Griffiths, 2002). Consequently other roles for rod-shaped cells and their advantages will have to be considered to support this point of view. Considerations will be postponed until after the mechanisms that likely functioned for the wall growth of Gram-positive and Gram-negative cells are briefly presented.

WALL GROWTH MECHANISMS FOR THE FIRST BACTERIUM

The strategies for both a Gram-positive and a Gram-negative cell presented above for cell enlargement have been studied both theoretically and experimentally. The one for Gram-positive cells is a well established whereas the one for Gram-negative bacteria is likely, but yet to be proven. It can be appreciated that both sidewall and pole formation for Gram-positive organisms are quite simple compared to the process that is used in the Gram-negative cell.

What are the possibilities for how saccular growth occurred in the first place? Could the suggestions presented above for the Gram-positive or Gram-negative forms of modern bacteria be those which functioned for the LUA cell that just had perfected methods to form a cross-linked fabric outside its cytoplasmic membrane? Being able to form a polymer outside the cytoplasmic membrane is very complex, and there must have been many problems to be overcome, but just the ability to lay down wall is not enough. When new murein is incorporated into the sacculus, what then? It is here that a whole additional set of quite different mechanisms had to be generated and had to come into play just before the time of the creation of domains of life. While the early mechanisms may not have been as sophisticated as those used in modern bacteria, at the start of the Bacterial domain they had to be sufficient, simple, and functional.

On the basis of the possibilities for only two major classes of cell, it would appear plausible that something like an "inside-to-outside" mechanism for Gram-positives would be much simpler than the "nona-muropeptide stretch" mechanism for Gram-negative cells. Moreover, it seems self-evident that a coccus instead of a rod-shaped organism is simpler and should have arisen earlier. So the first hunch would be that the original bacterium was a Gram-positive coccus. For division, this cell must have been capable of forming a septum or cross wall, possibly with a mechanism similar to that used for cross wall formation of *B. subtilis* or *Enterococcus hirae* in modern bacteria.

CONCLUSION ABOUT BACTERIAL ORIGINS

The first bacterial cell by definition had to have a sacculus. But other more specific thoughts about the first member of the Domain of Bacteria are more varied. It could have been a Gram-positive organism, that is, if Gram-positivity only denotes a thick peptidoglycan layer, Gupta, Siefert, and Fox apparently would agree. It was argued above that the Gram-positive strategy is much simpler than that for Gram-negative cell wall growth. Cavalier-Smith argues for a Gram-negative cell being first, although he mainly considers the cell membranes and not the murein wall and its role in bacterial growth.

During the time between the origin of the first cell and the first bacterium, the prokaryote cells had no murein layer. This is the likely possibility because the organisms only gradually become more successful and only as this happened would their turgor pressure increase. The simplest mode for Gram-negative growth so far suggested is the nona-muropeptide stretch model described above, which is much simpler than earlier models, but more complicated than the "inside-to-outside model" for Gram-positive bacteria. I find it difficult to imagine that the first cell, as Cavalier-Smith would postulate, was a Gram-negative cell (1987).

Of possible cell shapes, that of the coccus is simplest. If *E. hirae* were taken as the model for the first bacterium, such a cell would only need form a thick septum and bisect it and let physical forces do the rest. The semi-conservative process described above for *E. hirae* would not seem to be too difficult to implement with a few or no enzymes.

The reasons for believing that a rod-shaped organism came first have been cited. But in addition, there are biophysical arguments for the advantages of rod-shaped organisms over coccal type of cells based on kinetics of uptake of nutrients from the external environment. In many articles (many by myself) these issues have been discussed. There are several different cases. In an environment with significant concentrations of a resource, if every unit of surface area has the same concentration of uptake sites, the only design factor is the ratio of surface to volume. Smaller cells have higher ratios. For a fixed volume, thin rods or flat leaf-like structures have higher ratios, while spheres have the lowest ratios.

If different regions of the cell are different in the number of nutrient absorbing sites then the actual surface area of the cell is not the only quantity of significance, and in addition the number and kind of uptake systems per unit of surface area and the nature of the variation in amount from point to point is important.

On the other hand, when nutrients are present in very low concentration the factor of prime significance is the diffusion process from the bulk medium up to the cell surface. While the mathematics for different shaped objects is different, in actuality the effect of cell shape is quite minor at low concentration. Of course, even if the effect on growth rate is very slight, over many generations it could be important. The general conclusion from these considerations is that a rod shape is better than a coccoid shape for cells of the same volume.

Are there other advantages of rod-shaped growth over the growth of coccal type organisms? Two can be suggested. One is that the individual cells in a mass of cells in an aggregate may have a more effective exposure to the environment if it is in an assortment of randomly oriented rod-shaped cells than if it is in a compact mass of more spherical particles. This is a possibility that must be mathematically explored.

The second suggestion, and possibly more important one, is that a rod shape is safer when cell division is controlled in a certain way. Since the 1960s, it has been known for *E. coli* quite surely that at some definite time in the cell cycle the initiation of DNA replication event takes place. This in turn controls the timing of the subsequent cell division event. Other alternatives, in principle, are possible. This rod-shaped strategy may have been favored to preclude the possibility that the cell division event would interfere and break a chromosome. Consequently, a reason for assuming the rod-shape mode of cell growth could be that it is safer for the genetic integrity of the cell.

Contrast that with the behavior of cocci. For the case of *E. hirae*, it appears (Koch *et al.*, 1982; Koch and Higgins, 1984) that splitting of the septum is initiated when wall growth cannot occur fast enough for the cell's increase in cytoplasm. Because of the disparity of the ratio of volume growth rate to wall surface growth rate through out the cell cycle, the trigger could be turgor pressure, which inside the cell increases towards the end of septal closure. Such a system could err if the chromosome and the cell division cycle were not in proper temporal relationship and that would require an additional regulatory system that is not understood yet.

Relevant to this, Seifert and Fox (1998) observed that in the 16S rRNA evolutionary tree, there are several branches in which the cells switch from rod-shape to coccoid, but have not reverted during further evolutionary time. It can be imagined that the first bacterium was a rod to allow safe growth, but much later in the development of the bacteria clades with genetic deletions occurred to eliminate the sidewall region. This could have been dangerous unless new developments of various kinds occurred to control the chromosome cycle to be in accord with the cell division cycle.

This is an attractive possibility, but more sequence data for more species will be needed to check the mode of cell growth. However, for now, this is the best guess for the apparent initial role of rods and subsequent emergence of cocci.

Chapter 17

Diseases: Old and New

The pathogens infecting a widely distributed host species must either: (i) have an alternative host; (ii) be able to survive in a dormant state; or (iii) be non-destructive to their host. For pathogens of diploid hosts that have an obligatory sexual mode of reproduction, a particularly effective strategy is that of being sexually transmitted and the movement of the host allows the pathogen to disperse. There are two major problems with this STD type of strategy: First, a sexually transmitted disease must be, at least relatively, a "gentle pathogen" or a "prudent predator" and remain as such and not evolve towards greater virulence. The second, and quite important problem is that such a pathogen must, to a large extent, be able to avoid destroying its host's offspring. Both factors would seemingly be countermanded by the immediate selective advantage of being more virulent and producing more propagules of the pathogen. The first problem has been extensively discussed in the literature. A new proposal, presented here, is that mechanisms evolved to protect a previously infecting pathogen from other subsequent infecting pathogens and this may incidentally protect the host and its offspring. The second problem is related to the first and is a major factor with animal hosts. In large part, the "temperate" pathogens of hosts with immune systems actually depend on the host's immune system (and other protective systems) to keep the pathogens themselves from damaging the host too quickly. The lack of an effective immune system in the fetus (even with some maternal immunity) and in the early life of newborns implies that the persistence of STD pathogens may be favored by the limited protective systems of the fetus's mother to some degree. It is suggested that the mucosal immune system of the host may be able to reduce the danger of infection and of disease to the offspring. In the long run, this is favorable for the STD pathogen because then the young members of the host population can grow to become sexually mature. It is suggested here that the selection force driving both the generation of the pathogen to become "gentle" to its host and to limit the degree of infection of the fetus and neonate is the consequence of evolution of the pathogen to be able to inhibit other pathogens (even of the same species) from infecting the pathogen's host.

INTRODUCTION

This chapter is mainly about viruses in the Eukaryote human. This is because the relationship of host and pathogen are better known or at least thoroughly speculated about. However the relationship of the host and pathogen is the same no matter what the details of their positions in the tree of life and the molecular biological abilities of either.

Much host-pathogen biology, including that relevant to human disease, can be inferred through the study of viruses that infect bacteria, such as the bacteriophages T4 and Lambda, λ. These viruses have elaborate mechanisms to deal with their host, with themselves, with each other, and with other pathogens even of their own kind. The experimental elucidation of the elaborate molecular biological mechanisms of viruses to keep out other viruses have pointed a way that counters the evolutionary tendency of the pathogen to become more virulent to the host. An analysis of the molecular biological mechanisms and how they can function for the "good" of the host for bacterial viruses is being prepared and partially presented here. Such a model is to be contrasted with earlier models based on kin and on group selection (summarized in Anderson and May (1991) and Frank (1996). The paradigm presented here may well be quite general and it will be assumed that evolution can create and maintain a "gentle" pathogen in this way. According to this generalized model developed for viruses of prokaryotes, the lysogenic pathogen persists indefinitely in a temperate state in spite of the clear short-term selective advantage of changing to become virulent. This counter-intuitive behavior is accounted for by the existence of a positive selection mechanism functioning against other pathogens. Incidentally this leads to it being more "gentle" towards its host. I will use the term "gentle" in quotes to denote that the pathogen has a *quid pro quo* in the interaction with the host. Although proposed and justified for host-parasite relations in the virology of bacteria, I now suggest that this mechanism could function in host-parasite interactions no matter what organisms are involved. As the primary focus in this chapter, it will be presumed to function in the interactions of primates and retroviruses with as yet unknown specifics, and in particular to humans and the HIV virus.

Two topics will be considered: First, the problems that any pathogen would have surviving on a sparse, but clustered, host population, such as cave-dwelling primitive humans. The alternatives to developing the habit of being "gentle" to its host while being an STD are either: i) acquiring and using an ability to remain dormant for long periods of time; or ii) of also being able to infect an ubiquitously and commonly occurring alternative host species. These two possibilities will not be examined further here, but they do occur in nature. The option of being "gentle" to the host is the main topic here with respect to

STDs, but an important aspect of this is the second problem considered here; i.e., that an STD of a mammal would be favored if it prevents, to a considerable degree, the destruction of the neonates of its host.

THE ADVANTAGE OF BEING STD IN A SPARSELY POPULATED WORLD

There is an essential feature needed for an infectious disease of social animals to successfully disperse in small groups. This is a way to spread from one social group to another social group. In sundry diseases (Hoeprich *et al.*, 1994; Gilbert, 1993; Roizman, 1995; Mandell *et al.*, 1995), this is done by forming long-lived spores, or passing through intermediate hosts or vectors, or being carried by various animal and insect vectors over significant distances. Another very effective and quite successful way, however, is to develop a sexual mode of the transmission, which depends on the host actively interacting with the remote populations of his or her species. In general, many STD pathogens are not long lived in the environment outside the metazoan body and, in addition, are highly specific so that they do not usually have an alternate host. Consequently before the invention of the hypodermic needle, STDs usually had only a sexual mode of transmission within individuals of a species (Merrigan *et al.*, 1999). STDs must be extremely common since there are many that are quite mild (Hoeprich *et al.*, 1994) and I assume that there are many more that are so mild that they have not been detected.

Because of limited interaction between social groups, this strategy would necessarily be unsuccessful if the pathogen was highly lethal, since it would be eliminated by destruction of the host population. It can be presumed that the ebola virus infecting humans faced and regularly lost in this way in the past in rural Africa. Thus, it jumped from some animal into an isolated human social group, destroying that group and then itself, and thus this particular episode was ended. Any successful STD disease would need to be as “gentle” as possible while still retaining its infectiousness if it is to be effectively transmitted to other hosts during rare encounters between groups, especially when population levels are low and groups well separated.

A distinction needs to be made that differentiates survival of a “gentle pathogen” from evolution of the host by group selection. Group selection can only occur if there is the possibility that some host groups may contain a gene that confers some level of resistance to the pathogen. This gene would then be selected and become common and the descendants of these groups would become prominent.

The long-term persistence of a disease among sparsely distributed social animals will depend on the particulars of the social interaction between groups. Passage of an STD from group to group is aided by the social behavior of the host, which may institutionalize the transfer. For example, the social groups of many kinds of mammals center on a single dominant male or a female-bonded social group (Wrangham, 1980; Manson and Perry, 1993). Consequently, other (usually younger) males or females of a variety of species are ejected from their natal small groups. Sometimes an emigrating male may become the dominant individual in other groups or an emigrating female may enter other social groups to breed (and in either situation, incidentally carry STDs with them). In chimpanzee troops and other primates living in groups there is often the exchange of young females between groups as the females reach sexual maturity. Frequently, these females are forced to leave their original group. Possibly this same strategy was practiced by early human groups. After the female chimpanzees are ejected, they are taken into neighboring colonies. Not only is this mixing of populations known from direct observation of various non-human primate species, but it also can be deduced by the smaller degree of polymorphism for genetic markers exhibited in the male population of a clan of chimpanzees relative to the females in the same group.

This process may serve the primate species very well by limiting the effects of inbreeding. The explanation of these behaviors by geneticists and sociobiologists is that consanguinity is bad for any species (except obligatorily selfing-organisms, like Mendel's garden peas which do practice incest habitually). Whatever the validity of this explanation and how the custom arose, sexual mixing occurs between these social groups as an institutionalized process even in the absence of institutionalized prostitution, rape, and war. Unless these violent transmission events between host groups are fairly frequent within a small group, the behavior of exporting one sex decreases the inbreeding.

DISEASES IN PRIMITIVE HUMANS

A sexual transmission of disease from adult to adult gains most of the advantages of vertical transmission in not needing to be transmitted through the environment. Moreover and very importantly, it can survive in sparse clustered populations of hosts because of the sexual proclivities of the latter. In these two sentences the essential elements for a parasite (especially a virus) to survive by infecting humans as its only host in the Stone Age have been spelled out. In such circumstances it is necessary for a pathogen to be able to compensate for its host's low population density and its sporadic distribution. During the tail end of the most recent Ice Age, the human hosts survived in small, mostly isolated, groups by "chasing mastodons and other large game."

Those pathogens that depend on humans as a resource faced a much different problem after domestication of plants and animals than during the hunter/gatherer Stone Age, as discussed in the last paragraphs. By the time of the ancient Egyptian empire tuberculosis occurred in people, since by then the agriculture was effective, the population in cities became quite dense, and transmission through the air from person to person became efficient. The problem of pathogens became still quite different subsequent to the Industrial Revolution and still different in the medically sophisticated world of today. Possibly, at the time of Christ the world population was a thousand-fold larger than during the Ice Age, and probably, the world population of humans has increased ten thousand-fold since. Today, with the world human population greatly increased, with transportation easier, with higher local population densities, with more rapid migration, with mixing of humanity taking place at an unprecedented rate, and with medical advances, the situation has become entirely altered. These changes result in both a great increase and an alteration in the spectra of diseases. This has occurred particularly because these communicable diseases are transmitted effectively between people that are crowded closely. For this reason, it can be argued that although the major diseases of Stone Age humans were largely STDs (or with pathogens capable of remaining dormant or propagating in other hosts) the typical major infectious epidemic diseases experienced in the early Christian era and later were not STD. The aftermath of hypodermic needles, jet planes, and other modern inventions will be that the spectrum of diseases will become much different still in the future.

The two keys of the matter are: first, that many of today's relevant human STDs have had a long association with primates, including humans, and have had an opportunity to modify their hosts; and in return, the pathogens have been modified by their host's biology. Second, the same group-group interactions, as in the hunter/gatherer cultures, apply to other primate populations living in the wild today. So we can assume that STDs specializing in particular species have been around for a very long time in human and other primates, and that they should generally be "well-tuned to their host species", and particularly to its population density and sociality. Furthermore it can be assumed that only occasionally will such adapted STDs be dangerous or lethal to their specific host species, or at least to most of the individuals of that population. To the degree that HIV has recently entered the human population or more specifically the population of modern industrialized, medically treated, jet age men and women, it is now in an unfamiliar milieu and is especially dangerous (Gilbert, 1993).

THE SIMILARITY OF THE STRATEGY OF THE CAVE MAN'S DISEASES AND THE MODERN PRIMATE PATHOGENS

All STDs of cave men (I presume) and the STDs of today's wild primates face the same general problems. But here we will be more specific and consider the problems of the retroviral STDs of monkeys and man (Mandel *et al.*, 1996; Kurth and Norley, 1994). Now let us match the retroviruses of primates to the design criteria for an STD pathogen of sparse clustered populations. Point by point these criteria are met by viruses abundant in non-human primate populations, such as SIV that is resident in the African Green Monkey. Of course, SIV may be more virulent when transferred to different monkeys; e.g., to the geographically distant Asian ones, than the species from which it was initially isolated. Of course, SIV would be virulent in animals that happen to have a defective immune system. The general point is that the small deviation from the behavior optimized for a retroviruses' life in its native host now leads to catastrophic problems in the variant host, whether it is the human, the Asian monkey, or the immune compromised host.

It is assumed that HIV emigrated to humans relatively recently (see Eigen and Nieselt-Struwe, 1990; Reines, 1966) and it and the new human hosts have not had a chance to adapt genetically to each other, though there are some indications that there have been developments in this direction (Ewald, 1994). One of the maladaptations between the disease and its relatively new human host, which is a very clear example of a factor affecting disease virulence, is that the human immune system after HIV infection deteriorates in 5 to 15 years. This same period would be unimportant to non-human primate populations in the wild because they have shorter mean life spans and maturation periods and may generally die of other causes before the loss of their immune system. On this basis, it may be that the changes needed to re-establish the "gentle" parasite mode in the new human hosts are minimal: For example, just a shortening of the human life span to the life span at the turn of the century level would make the disease relatively unimportant. I hope, of course, that this is not an option that will occur.

A disturbing, but realistic, suggestion of a change that would certainly make HIV infection relatively "gentle" is the re-emergence of life-shortening infectious diseases, such as tuberculosis. There are many other diseases that may erupt as the antibiotic era closes, and with the loss of efficacy of many antibiotics, the human life span may decrease dramatically and the immune system outlast the shortened length of life due to deaths of individuals from other infectious diseases. Then there would be a lower proportion of individuals with signs of

ARC or AIDS. Changes of the other kinds will be suggested below that might increase the longevity of the immune system of an AIDS-infected individual in a long-lived population, but others might decrease it.

SELECTION FOR "GENTLE" RETROVIRUSES AND THE MUCOSAL IMMUNE SYSTEM

Of the many kinds of immunological responses, the ones that function at mucosal surfaces are most relevant to STDs during their transmission from individual to individual (Testacy *et al.*, 1988, 2003; Quesada-Rolander *et al.*, 1996; Overbaugh *et al.*, 1996; Edwards *et al.*, 1996; Van Rompay *et al.*, 1996). An additional feature relevant to STDs epidemiology is that the sexual tract is in the right place for T cells and antibodies to prevent new infections.

The argument extended here from the prokaryotic cases (Koch, 2006) is that positive selection would lead to the development of mechanisms under the control of the virus to exclude a superinfecting pathogen. The model as deduced from the study of bacteriophages is that viral mechanisms that prevent superinfection by other pathogens in nature must have been under positive selection. With such mechanisms the virus would not be overtaken by other pathogens, and, therefore, antiviral activity of certain kinds (even towards the infecting viruses of the same kind) would be in the immediate self-interest of the originally infecting virus. If the first arriving pathogen is directly selected for its ability to keep out other viruses of as many kinds as possible, it is well served to establish such an operating block, even if the newly arriving pathogen is genetically identical to itself.

Translating this argument to the case of a virus of a metazoan host, the situation that corresponds to preventing a new infection of the multicellular host organism is preventing entry of the STD agent into the tissues of the host's sexual tract. It would appear that the most effective way for this circumstance to be implemented by an infecting pathogen is to stimulate the host to mount both a cellular and humoral (especially IgA) response in the mucosal surfaces against the virus. This need not have any major influence on the pathogen particles already inside other parts of the body, but some minor side effects might well ameliorate or slow down the disease process within the host. Extensive studies of mucosal immune responses have and are being made. While these studies were/are largely designed to look for opportunities for future vaccine development, they serve for the present purposes of showing that IgA and T-cell responses are made in mucosal tissues (Mestecky and Russell, 2003).

On this basis I argue here that the retroviruses (and other STDs) may have been selected for their ability to instigate or stimulate a mucosal immune response. Unfortunately, I have not been able to find experimental data in the literature to test this hypothesis. This is not because there has been lack of interest in this question, but because it would be hard to decide whether an infected person or organism had been subsequently infected with a different kind of the STD species. On the other hand, work has been on-going for some time to develop a mucosal approach to HIV vaccine development (Kutteh *et al.*, 1988; Forrest 1992) and this may incidentally lead to an answer.

A strong indicator that HIV is usually prevented from infecting children perinatally is that, while the majority of children born to HIV-positive mothers are not infected, they almost all carry maternal antibodies against HIV. Presumably they had them *in utero*.

Going even farther out on a limb, one can imagine that the mucosal immune response could protect the fetus and the baby during the birthing process, and consequently, the neonate may be disease free. This might be accomplished by a strong IgA or T-cell response, or by an IgG response that is delivered through the placenta into the fetus. If it were possible in terms of the host's immunological repertoire modulated by some stimulatory action of the pathogen, this would increase the fitness of the STD. Such a situation would lead to continued selection for such genetic variants of the host and in the STD through successive generations of virus and host in an environment with high levels of alternative, but similar, pathogens. These developments would include, of course, changes that would "tune" the growth of the retrovirus to a particular primate species. These changes may lead to effective prevention of infection by secondary pathogens or superinfection of the resident sexual disease by other copies of itself, but would have a long term and incidental effect of maintaining the host population. Such selection would lead to corresponding changes so that the pathogen becomes milder to the host as a consequence of induction of greater immunogenicity against the resident pathogen and beneficial to the host because the first pathogen would prevent other, possibly more dangerous, pathogens from entering and destroying the host.

The converse possibility is that the pathogen causes deleterious effects due to stimulation of the immune system. In many cases these effects are significant, but generally not lethal.

SECRETORY LEUKOCYTE PROTEASE INHIBITOR

The role of secretory leukocyte protease inhibitor (SLPI) may also be a factor in HIV infection. It is known that salivary gland tissue may have a role in suppressing transmission by the oral route. Wahl *et al.*, (1997) were able to account for the rarity of oral transmission even though there is HIV in oral secretions. Once inside the oral cavity, HIV1 exposure to antiviral levels of SLPI, their data suggest, this may impede infection. An extrapolation for which there is no data could be that HIV might stimulate the synthesis of SLPI. A further extension is that the same effect might take place in the genital tract and prevent secondary infection of HIV viruses or variants or a broader class of STDs. These suggestions could be tested.

VARIABLE STIMULATION OF THE IMMUNE SYSTEM OF PRIMATES UNDER DIFFERENT CONDITIONS

The long-term survival of the pathogen in a vertebrate host depends on proper balance (or evasion) of the pathogen with the host's immune system. A possibly critical factor of difference among modern humans, the Stone Age man, and the modern non-human primate needs to be recognized here. Some of the possible differences would make the response greater or weaker, and could consequently support or conflict with the hypothesis proposed here. This balance could depend on the variety and extent of immune stimulation in general and is in addition to the responses of these host species to their particular retroviruses. The stimulation of the immune system, variety of stimulations, and the age dependency of individuals exposed to other antigens before they are challenged by viral antigen do vary greatly (see for example, Miedema and Klein, 1996). There must be marked differences in the antigenic stimulation of the members of primate societies in the wild, the humans in developing countries, and the humans in developed countries. Either too much, too little, or the wrong kind of stimulation may mean that the immune system, particularly the mucosal part of the system, will be either over or under active. A slight variation may eliminate the pathogen or fail to prevent new pathogens from entering and supplanting the original one. The immune system may be secondarily modified by destruction of certain specialized T-cells which respond to the HIV or SIV pathogens. As a speculative example, female chimpanzees engage in some sexual activity for some time before they become pregnant. If the female has been infected with SIV for several years before she becomes fecund, it would not be surprising if she had slowly developed a range of mucosal immune responses that would have a protective effect on her developing fetus. This effect might not occur if she had been infected just prior to becoming pregnant. Some aspects of the AIDS disease process in humans, it is well known, are similar to autoimmune diseases and the destruction of cells of the immune system bearing viral or similar receptors. All that is being suggested here is that such phenomena need be only slightly altered in the not quite yet-adapted HIV-human pair from those of naturally adapted retrovirus-primate pairs to give virulent disease. In the current world AIDS epidemic these could affect timing of symptom onset and the severity of debilitating disease.

At an earlier time when our blood supplies were contaminated, hemophiliacs were very likely to receive the AIDS virus and become HIV seropositive. A significant point is that their time to conversion to ARC and to fulminating AIDS was 90% longer than other non-hemophiliac seropositive persons at that time (Darby *et al.*, 1995). One possible reason is that the medically treated hemophiliacs were being continuously challenged and immunized against a large variety of other substances that were present in the various blood transfusions that they regularly received. An additional factor concerning the AIDS disease in the past was that when hemophilia patients were initially infected by transfusion, the disease would have been started with a much larger number of viruses than that transmitted by usual or unusual sexual practices or by a drug addict's re-used needle. Consequently, their immune system was stimulated in a way that the immune systems of normal people infected by sexual contact are not, simply because the intensity of the immune challenge for the hemophiliac was greater. These possibilities raise the suggestion that African green monkeys are immunologically equivalent to human hemophiliacs receiving blood transfusions. As a result, an African green monkey from which SIV can be isolated might have an especially effective immune system and be able to continuously destroy a much larger proportion of retroviruses and thus limit the viremia. This could mean that it would live for a long time before it would become immunologically deficient. In any case, it will be interesting to see the disease progression in fresh hemophiliac patients that were infected by sexual transmission and not any more by transfusions now that they receive recombinant clotting factors without these multiple sources of diverse immunogenic stimulation or of HIV. This could be a test of the proposed theory.

Immune reactivity can depend on the presence of other pathogens. Infection with mycoplasma, Herpes, Epstein-Barr, and several other viruses may affect how HIV infection leads to the immune deficiency state. With co-infection of such viruses, it is likely that the time at which AIDS erupts may be sped or slowed in either the human or monkey. This raises the possibility that a different spectrum of viruses and other ubiquitous diseases of a species of a host might influence the course of a retroviral disease. A sometimes-symptomless retrovirus infection that leads to the development of a debilitating immune deficiency might do so differently when infecting a different host species. On the contrary, other pathogens may trigger the emergence of the AIDS virus from the host's chromosomes. These assorted immunological events could greatly affect the life history of a retrovirus in any new primate host. Consequently, it can be argued that the AIDS virus may well have been adapted to be "gentle" and virtually unobstructive in an old host, but at present in humans, those strategies do not work the same because the human pathogenicity is different. Because of our social organization, our life expectancy, and our antigenic environment, the outcome of this parasitism could well be quite different. Possibly the finding is relevant to the case of an individual that became infected with only one strain even though two different varieties of HIV were transfused simultaneously into him (Diaz *et al.*, 1996).

A very striking observation was made several years ago (see Mitler, Antia, and Levin 1996) and considers the significance of the two experimental publications. The major conclusion is that the HIV viruses grow very rapidly and are very rapidly destroyed by the apparently healthy, but HIV positive, individuals. It leaves unanswered how much of this destruction is due directly to the action of the host versus the action of the host as modified by the virus. However, there is the possibility that to some degree this is due to mechanisms implemented by the resident HIV virus.

EPIDEMIOLOGY OF AIDS AND THE "RAPID" CHANGE OF HIV

By comparing the epidemiology of HIV-I and HIV-II in their respective parts of Africa, Ewald (1994) has made the case that the diseases vary in the virulent/gentle scale in a way that is correlated with the number of sexual partners and the group's societal norms. This would be predicted by a game theory type calculation on the assumption that the virus was omniscient without any suggestion of how it could arrange such things. Accepting his ideas and facts as true, we can only be amazed by the speed with which the virus can apparently change its strategy and optimize the course of infection to the new conditions presented by the new host. I think that this means something much more significant than the point made by Ewald. As an alternative to *de novo* evolution, I suggest that this apparent rapid adaptation is consistent with the hypothesis that STD retroviruses have been exposed over long evolutionary times (in terms of millions of years) to fluctuations in the behaviors of a series of hosts (or a host under a series of very different conditions) and have developed and retained a genetic repertoire to be able to achieve slight functional variations from the molecular point of view to allow them to track and respond to the behavior of their current host. Although the genetic sequences are now known, the roles of all the gene products are not fully evident and some of the regulatory genes may have functions over a long time period of years and centuries.

WHY THE MAMMALIAN STD PATHOGENS MUST PROTECT THE HOST'S NEWBORN FROM THEIR OWN LETHAL ACTION

A balance must be maintained for long-term existence between the virus spreading from host to host and in its injuring or killing the children of the infected host. The biology of pediatric AIDS is reviewed in Pizzo and Wilfert (1998), Roizman (1995), and Sweet and Gibbs (1995). Teleologically, the young of the host's must be spared to be used as a resource in the future, but that necessary truth provides no mechanistic way to implement such a situation. HTLV-I (see below) seems to have achieved this balance by infecting most offspring of infected mothers and then by growing so slowly that the children's life before puberty is virtually unaffected. Such slow growth permits propagation of the virus to virus-free individuals within a society in which sexual contact is frequent and avoids the usual destructive effects of vertical transmission from mother to child. The AIDS viruses, HIV-I and HIV-II, do not seem able to avoid injuring the host's children when they become infected. But still a large proportion of the young of infected mothers do not become infected—more than 60% and in some estimates as much as 85% for HIV-1. Some infants, born infected, eliminate the HIV virus (Roques *et al.*, 1995).

An infectious disease spreading to the next generation by vertical transmission has advantages over pathogens disseminated in other ways. The main advantage is that the disease organism never has to survive in the environment outside of the host. Survival of the pathogen outside of its host, say, in air or water, takes special protective mechanisms. Although the fetus is a convenient and a built-in susceptible host, the vertical process in vertebrates has a special complexity beyond that present in the cases when the host grows by binary fission as bacteria do. This is because the animal host gives birth to an immunologically ill-equipped neonate that may not be able to survive due to damage that could be caused by the pathogen. This is a critical problem for the STD survival strategy because they generally have no alternative propagation strategy such as survival in alternative hosts or in the environment.

Because the newborn offspring do not yet have a full immunity system the result is often catastrophic. This is what happens when some pathogens, such as Herpes Type II, Rubella, or Cytomegalovirus infect an embryo, a fetus, or a newborn child. In these cases, because the offspring are without a fully protective immune mechanism, the infection may be destructive or lethal to the offspring and is much more severe than if the pathogen infects an adult that has a responsive immune system and has an immune repertoire already in place.

Therefore, in many cases a human pathogen cannot effectively persist via pure vertical transmission simply because the pathogen "depends" on the presence of a host immune response system to bridle its growth. Viruses like Herpes, Rubella, and Cytomegalovirus that go across uterine, vaginal, and placental tissue are frequently dangerous to the fetus, but of course these viruses survive because their primary means of spreading through the population is from adult individual to adult individual and are not dependent on vertical transmission for their survival. Destroying a few children apparently does not upset the growth success of these viruses. However, from the viewpoint of a disease that has elected to only use the non-virulent STD strategy, going directly from mother to baby is a destructive strategy even though it is expedient and does have advantages. This strategy must somehow be avoided, and it apparently is often curbed.

There may be a number of factors involved. One is that the AIDS virus is just not very infectious or long lived. Some suggestion of this comes from the fact that AIDS is not transmitted by way of bloodsucking mosquitoes or efficiently through punctures of health care workers with contaminated needles. Presumably, in the mosquito example, this is because the virus does not last long enough between successive blood meals of the female mosquito. Additionally, it may not be transmitted because infection requires a large inoculum. Similarly for the second example, the fact that HIV is quite seldom transmitted to health care workers through needle sticks is possibly because most particles are inactive and many viable particles are needed to cause infection in this manner.

While these could be trivial or based on molecular biological necessity, I suggest that the virus has been selected to extend these characters. Evidently, this is the characteristic of particular viruses, but it also may be a human or primate characteristic. This can be if mankind has habitually, over the eons, been exposed to retrovirus type pathogens. For both virus and host their fitness is increased in not permitting infection of the still immunological incompetent neonates. It may be that the low infectivity of neonates is because of some molecular biological or biochemical limitation or necessity, but in the retrovirus case this can be presumed to have been particularly applicable. The argument is that there are some diseases that are transmitted very efficiently by mosquito bites and by limited blood-to-blood contact (malaria and Hepatitis B jump to mind). So I suggest that the retrovirus is poorly infective due to innate biological mechanisms evolved to favor the long-term goals of survival as an STD. This is a prediction of the model proposed here that would have aspects that could be experimentally pursued.

Although I will discuss below how the retrovirus Human T-cell lymphotrophic virus I (HTLV-1) avoids this difficulty by known mechanisms, just how HIV perinatally infects only a portion (15–30%) of children (Scott, 1994;

Cotton, 1994) and not a larger percentage is not clear. Women that acquire HIV after delivery have a higher transmission rate to their children by breast-feeding than do women previously infected. We have also a hint because of the finding that reducing the viremia by zidovudine (AZT) treatment before and during the delivery process reduces the transmission to the child (Connor *et al.*, 1995; Spector *et al.*, 1994; Rouse *et al.*, 1995; Brossard *et al.*, 1995.). However, there are now several reports that children can be initially infected and apparently clear themselves of the infection (Bryson, 1995); this makes it difficult to compare different studies.

THE COPING STRATEGIES OF HIV, HTLV-I, AND HTLV-II

HTLV-I and HTLV-II are both retroviruses, of the subclass oncornaviruses that propagate primarily in human T lymphocytes. The former parasitizes CD4-bearing helper T cells and the latter, CD8-bearing cytotoxic T cells. (Höllsberg and Hafler, 1996). HIV is a retrovirus of the subclass, lentivirus, and both it and the oncornaviruses incorporate their reverse transcribed double-stranded DNA into a random chromosome of a human cell as the heart of their survival strategy. Both classes of virus infect only a particular human cell-type and this one must be one that continues to divide. Nerve cells and kidney cells are not appropriate host cells because in the adult they almost never divide and therefore the equivalent of the prophage state would never be created or propagated. Another inappropriate cell type would be a rapidly growing type type such as epithelial cells, because the stem cells remain, but the other cell is sloughed from the skin or into the intestine and does not remain and propagate within the body.

The oncornaviruses, but not the lentiviruses, have a pattern that allows both effective vertical transmission and horizontal transmission to new hosts. The viruses are passed vertically to the neonate, although not with high efficiency and not in a way that causes childhood death. They are also passed as an STD between sexually active individuals. In addition, in the modern world they can also be passed via needles. Before extensive movements and mixing of the peoples of the world population, both HTLV viruses were geographically restricted in distribution. Although HTLV-I is now found world wide, it was and is highly abundant in southern Japan and the Caribbean. It is also present in South America, West and Central Africa, India, Melanesia, and Iran. There is evidence that similar viruses inhabit non-human primates. HTLV-II is now highly prevalent in intravenous drug users, but is (and presumably was) abundant in the Guaymi tribe in Panama and, more generally, in Amer-Indians.

The strategy of both HTLV viruses in the absence of intravenous drug usage and rapid movements of peoples was very close to the "gentle" pathogen persuasion. Both viruses persist with little damage to their host by replicating very slowly while being passed vertically or between adults as an STD. Being poorly transmitted from cell to cell and individual to individual favors the "gentle" character, but another feature is that they have mechanisms to cause the T-cells to replicate. In AIDS, depletion of T-cells in long-term HIV infection occurs, but this does not occur in HTLV. This allows more opportunities for the virus to be transmitted between humans. These viruses are poorly transmissible and transmission from male to female is rare and the transmission from female to male is very slow (at least for HTLV-I). Though it is passed presumably in the same three ways that HIV is passed to the neonate; i.e., *in utero*, perinatally by blood-to-blood transfer, and (predominately) postnatally via mother's milk, there is little damage to the child because the virus grows so slowly. Thus the immune system has time to develop and respond to these viruses and is even aided by the increased growth of T-cells. Although both viruses are almost perfect "gentle" pathogens, the current longevity of the human host is sufficient to cause problems. For example, there is the occasional generation by HTLV-I of an adult T-cell leukemia (ATL). Less frequently T-cell chronic lymphocytic leukemia, non-Hodgkin's lymphoma, mycosis fungicides, and Sezary syndrome arise. HTLV-II has not been adequately studied, but appears to be even more "gentle", and only very rarely causes glomerular nephritis and HTLV-associated myelopathy.

Molecular biological information is mainly available for HTLV-I, but the studies of the *rex* and *tax* genes are relevant to the trick that the virus has of alternately stimulating T-cell growth and then subsequently switching to HTLV-I replication mode (Yoshida *et al.*, 1991). Of particular importance to the major thesis of this speculative chapter is that the *rex* gene of HTLV-I interacts with the *rev* gene of HIV and this may suggest that the human virus, like those of the bacterial/bacterial viruses discussed in a paper by Koch A.L. 2006. Evolution of Temperate Pathogens: The Bacteriophage/Bacteria Paradigm FEMS Micro Review (in press), may act to protect their host against another pathogen.

HOW HIV IS "GENTLE" TO ITS HOST

Some of the methods that HIV uses to be lysogenic and spread from host to host are obvious from its generally known biology. Most obvious is that as a retrovirus, in principle, it would only be able to grow as the provirus in concert with a reproducing cell. HIV for example, can enter a quiescent cell and give rise to DNA products, but the DNA is only integrated after activation of the cell and replication of its chromosome. But the sole use of CD4 as a target receptor limits its opportunity to grow and reproduce. Maintenance of the latent state is another essential factor that keeps the virus from rapidly destroying the host's immune system. This is in addition to the host's immune system as an essential safeguard against viral long-term propagation within its current host and lessens transmission to other hosts.

SOME GENERAL CONCLUSIONS ABOUT HOST-PARASITE RELATIONS

(i) The evolutionary mechanisms that maintain the non-virulent state of a pathogen are of high interest. Since there are a large number of mild pathogens extent and in spite of the selective advantage that a virulent mutant would have in the short term, the "gentle" pathogen state does arise and does persist. A proposed mechanism for this, not involving group or kin selection, has been made above (and by Koch, 2006. Evolution of Temperate Pathogens: The Bacteriophage/Bacteria Paradigm FEMS Micro Review (in press). This new alternative model to sociobiological approaches not only could account for the selection of "gentle" behavior, but also would cause this state to be stable because it results from positive selection. This new model is that the pathogen protects itself (and incidentally its host) against other similar or identical pathogens arriving subsequently and that this generates and maintains the "gentle" state of the first infecting pathogens.

(ii) It is argued that pathogens that depended on humans as a resource faced a much different problem during the hunter/gatherer Stone Age, where the spectrum of pathogens could be expected to have been largely STDs because of the sparse and clustered availability of human hosts. Almost of necessity they must have been "gentle" because the interactions between social groups must have been infrequent. Before moving to humans, HIV must have been gentle to its non-human primate host, and possibly HIV has not had time to readjust to its current host.

(iii) Being a "gentle" pathogen requires elaborate controls to self-limit growth. Persistence of a sexually transmitted disease depends, therefore, on growth inhibitory mechanisms in part coded by the virus, but frequently dependent on host function. The host immune system that limits the viremia and viral encoded mechanisms may act to modulate the immune response and act in other ways to control and limit growth. It can be assumed that the lentiviruses of non-human primates, such as SIV (Simian Immunodeficiency Virus), are adapted to a low rate of vertical transmission because of the devastating action of many viruses on neonates due to the latter's underdeveloped immune system. HIV is transmitted to offspring *in utero*, perinatally, or via breast-feeding, but the transmission is less efficient than for some other viral diseases. This suggests that either or both host and viral mechanisms restrict vertical transmission or its effects. In the case of HIV, this restriction was probably not as effective as in the lentiviruses of non-human primates that probably resided for a long period of time within a species in the wild.

(iv) The retrovirus strategy depends on selectively infecting a restricted class of cells, mainly the CD4+ or CD8+ T helper cells. These and other T and B cells happen to be nearly the only suitable cells *a priori* because these continue to grow, replicate, and divide in the adult; and moreover their progeny cells remain inside the adult host. Such cells are the necessary condition for the strategy of retroviral growth in the provirus state.

(v) It is proposed here that the response of the mucosal part of the host immune system is the key factor to the prevention of secondary viral infection by even the same species and also to prevent the infection of the offspring of an infected female. This is because the selection for an ability to elicit an effective mucosal immune response by products generated by the virus would block secondary infection and therefore is in the best interests of the first infecting virus. Importantly, it also makes the resident pathogen less destructive to its host and its host's progeny. I suggest that the role of mucosal immunological response is critical to the biology of the STD retroviruses.

(vi) The important link in the STD lifestyle as typified by the HIV/human interaction is that the virus works against itself under conditions in which the hosts are at a premium and it must somehow protect the fetus and neonate. It is the fact that such protection is manifest even with HIV infection of humans. In the absence of medical treatment about 85% of the offspring of HIV-infected mother are not HIV infected. This number may be closer to 100% with simian viruses in wild monkeys. Compared with some other viral diseases, this suggests that specific immune protection of the fetus perinatally does occur.

(vii) The change from the forebears of HIV, presumed to be a "gentle" virus of its non-human primates, to the devastating human virus of AIDS is mostly because the new host of the virus lives much longer than its previous host. Added to this are the factors due to the human host's social behavior, most notably the extensive movements of humans from place to place. Also HIV appears very ungentle now because of its spread to a greatly expanded habitat, created first by homosexual involvement and intravenous drug use, and now simply because of the promiscuity of humans.

(viii) When a well-organized military force prepares to engage in a battle or campaign, the first things that the strategists consider are the opponents' goals, abilities, resources, and likely strategies and tactics. Similarly, to understand the essential features necessary to counter the current AIDS epidemic, it is crucial for the health profession to consider the situation from the virus's point of view. With all the AIDS research that has occurred, too little of this type of contemplation has been done. Consider: How was HIV well adapted in its ancestral host population? How is AIDS not yet fully in tune with its newest host, *Homo sapiens*? Can HIV protect itself against other viruses, including exogenous HIV itself? Does it implement mechanisms to decrease the risk of infection to the host's offspring?

Part 4

Antibiosis

Chapter 18

Lysozymes as Alternatives to β-lactams Antibiotics Acting on the Bacterial Wall

For a bacterium to enlarge its sacculus, there are many specialized steps that are not part of the growth strategies of plant or animal cells. Any of the biochemical steps unique to bacteria are potential targets that could be attacked by higher organisms and interfere or kill bacteria without harm to themselves. Such attacking agents would have no affect on non-bacterial organisms that produce the antibiotic substance. So antagonists have been developed over the last two billion years and are abundant in animals, plants, fungi, protozoa, and even a few bacteria. Here we briefly review the β-lactams, and then consider the lysozymes and their biology and compare the two.

THE β-LACTAMS

There are two bacterial targets that are especially at risk for attack. The most ideal target for interference is the bond that forms the tail-to-tail cross-links between two penta-muropeptides. This tail-to-tail bond cross-links the muropeptide, and therefore, has been discussed at length in this book. This target is particularly important since man has exploited β-lactam type antibiotics in a massive way and added variations to these chemicals in the last seventy years. This target has been particularly favorable because the tail-to-tail peptide bond in murein is unique to bacteria. This uniqueness arises of necessity because the tail-to-tail bond is formed outside the cytoplasm membrane and cannot utilize the usual cellular biochemical synthetic processes. Thus, its synthesis has to be different than that for almost all other peptide (amide) bonds. To emphasize what has been discussed in the earlier chapters, the penta-muropeptides bring their own sources of energy. Part of that energy is an extra D-Alanine peptide bond supplied in the disaccharide penta-muropeptide units and another part is from the connection made in the inside of the cell to the bactoprenol lipid. This is the agent that allows the wall unit to be extruded. Special transpeptidases (a PBP) covalently insert the disaccharide penta-muropeptide into the wall. These PBPs, although they have a transpeptidase activity, are special in that they have no hydrolytic activity and can only function as a transpeptidase. These enzymes have a structure such that water cannot get into the active sites of the functioning enzyme. It is this feature that allows them to synthesize murein. The β-lactam type antibiotics act irreversibly to inactivate these transpeptidases (endopeptidases).

THE LYSOZYMES

Another saccular target class is attacked by the "lysozymes." Their target is a specific glycosyl linkage in the glycan chains of murein. These glycosyl bonds are between the hexoses NAG and NAM. The destruction of this target is not achieved with an antibiotic chemical, but rather with a small hydrolytic enzyme. Sir Alexander Fleming first detected a member of this lysozyme class in 1921 (Fleming, 1922) as a "remarkable bacteriolytic element." There are many variants and variations of lysozyme in nature. They are present in birds, phages, bacteria, fungi, invertebrates (insects), and plants. In humans, they are present in tears, saliva, nasal mucus, leucocytes, skin, fingernails, and milk. The lysozyme of hen egg white is abundant, small, easily purified, and easily crystallized, and has served an important role in biochemistry and in immunology. They cleave between the C-1 of NAM and the C-4 of NAG. There have been two books devoted to the lysozymes (Osserman, Canfield, and Beychok. 1972; Jollès, 1996).

What is so extra special about the lysozymes? The initial answer has to do with their antibacterial properties protecting the eukaryotes against bacteria. But this is only part of the story because the plant lysozyme molecule has a connection with chitinase which breaks down the wall of fungi and insects. There is an evolutionary connection with the α-lactalbumins of avian eggs that are a major food source for the developing bird. It and a variety of lysozymes together bind the calcium needed by the developing bird.

The most studied group of lysozymes is that of the lysozyme C superfamily. It can be assumed that the superfamily originated 600 million years ago as an antibacterial agent, but subsequently was optimized to serve a role for nutrition to the mammal by aiding in the digestion of bacteria. For the avian it serves for protection of the fertilized egg against bacteria and favors the production of the next generation. Initially, the C designated the chicken egg- white form, but molecules with quite similar sequences are found in other birds, primates, artiodactyls, other mammals, turtles, fish, insects, monotremes. There are other families; e.g., Lysozyme G and Lysozyme F which are quite different. Phage lysozymes serve a different role (namely to allow the phage to exit from the bacterial cell where it grew) and thus are vital for the bacteria virus way of life. They are classed in several groups: V, G, λ and the CH. The first three have some relationship with the C type superfamily. The CH form is related to the *Streptomyces erythraeus* lysozyme.

All the plant lysozymes have chitinase activity. This is presumably because the chitin-encased fungi are the major predatory organisms of plant tissues. While the cuticle of the leaf structure can effectively keep at least most bacteria out, the lysozyme will protect the plant tissue against this fungal danger. Insects use lysozymes for protection and for nutrition. This is especially so in flies.

THE FUTURE

If we cannot make lactams more secure, there may be ways to make use of lysozymes for medical purposes more functional. This might involve reducing the size (already small) of the enzyme, humanizing it so that it is not immunogenic, making use of its already known resistance to heat inactivation, and applying it to human infections. There is no reason not to pursue this aim, particularly if bacteria become more resistant to chemical structures.

Chapter 19

Development of Wall Antibioics and Bacterial Counter-Measures

The bond that cross-links two muropeptides together in the growing sacculus is a special amide bond called a tail-to-tail linkage. Such a bond is also called an endopeptide. Consequently, the transpeptidase that forms it and the hydrolyases (autolysins) that rupture it have special features that distinguish them from enzymes dealing with the usual head-to-tail peptide bonds. For antibiosis an important feature is that the conformation of the active site in the transpeptidase has a structure complimentary both to the tail-to-tail bond and to the family of four-membered β-lactam ring antibiotics. Because of this similarity, the β-lactam ring binds to the active site of the transpeptidase. The β-lactam ring is more strained than if it had a large number of atoms, because the bond angles present in this small ring are smaller than usual. Consequently it reacts and inactivates the transpeptidase rapidly and irreversibly. This is the energetic feature of the structure of β-lactams, which favor its covalent reaction with the active site of the transpeptidase. β-lactam binding blocks cell wall growth and is the basis of penicillin action.

This is the beginning of the story. The evolution of β-lactam production allowed organisms, including some bacteria, to destroy other bacteria. Later, bacteria evolved β-lactamases to protect themselves. Still later, additional β-lactams evolved. Still later, humans developed the ability to use penicillins and cephalosporins as antibiotics. Bacteria developed counter-measures to many of these antibiotics, both before and after the pharmaceutical industry entered the picture. The battle of bacteria versus the medical community is continuing.

PRODUCTION OF β-LACTAMS AND SUBSEQUENT PRODUCTION OF β-LACTAMASES

After the initial evolution of Bacteria, Archaea, and Eukarya there was a great explosion in biodiversity. For their own benefit, some organisms developed chemical agents to harm other organisms. These classes of agents included: phytoalexins, antibiotics, lysozymes, proteases, immunoglobulins, etc. From this list, the antibiotics have been the most amenable for medical usage. The important fact is that the main evolution of the antibiotics preceded the success of the pharmaceutical industry in β-lactam manufacture by a billion years.

β-LACTAMS

The β-lactams, because of their structural similarity to the tail-to-tail bond in murein (Figure 19.1, lower structure), bind to the endotranspeptidase enzymes that carry out the tail-to-tail cross-link formation and irreversibly inactivate them. The basic biosynthetic pathway for β-lactams was evolved in the Streptomycetes. This first penicillin derivative or a very similar antibiotic (Koch, 2001) was probably iso-penicillin N. Figure 19.2 shows the metabolic scheme for β-lactam synthesis. The pathway was subsequently transferred into the fungal species *Penicillium, Aspergillus*, and *Cephalosporium*, and thereafter was

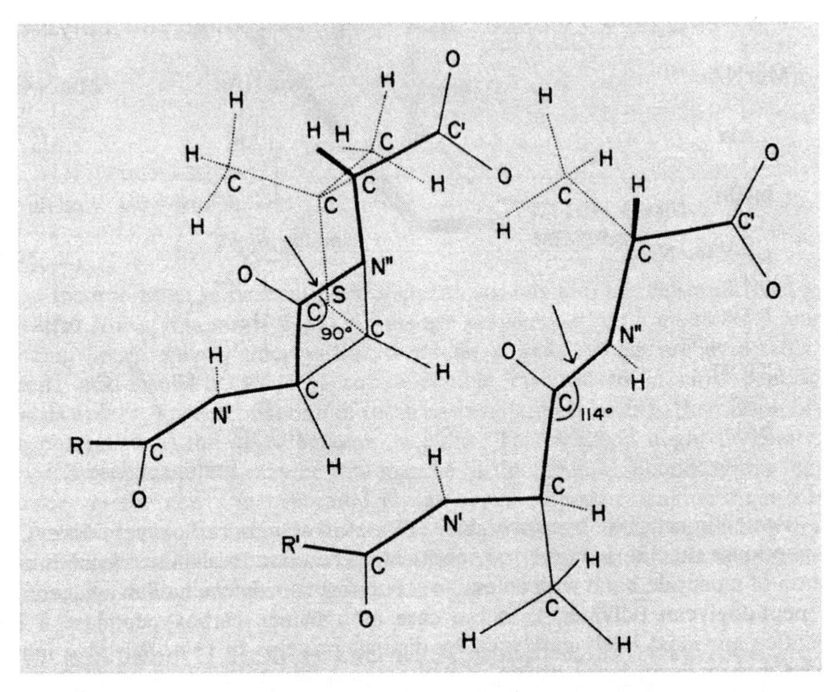

Figure 19.1 The structural stereo-formulae for acyl-D-alanyl-D-alanine (lower) and penicillin (upper). Figure taken from Waxman and Strominger (1982). The similarity between these two structures was the first evidence that β-lactam molecules can enter the active site in the transpeptidase and bind to it. Because of the bond stress in the 4-membered ring, it can covalently react and inactivate the enzyme, resulting in blocked saccular growth and lysis. The arrows indicate the lactam bonds, which are similar to peptide or amide bonds.

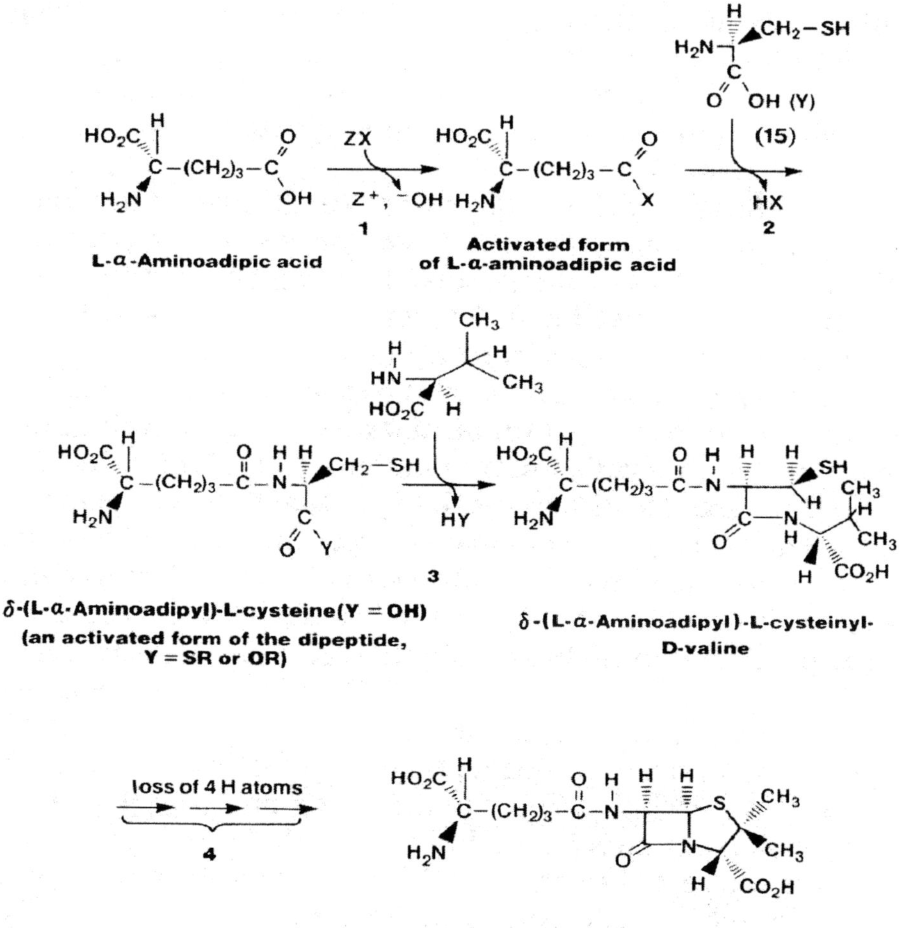

Figure 19.2 Biosynthesis of iso-penicillin N. Taken from Queener and Neuss (1982).

modified and the product diversified (Figure 19.3). In the vicinity of a producing organism the outpouring of β-lactams could inhibit the growth of susceptible bacterial neighbors or even kill them so that their contents could be utilized.

Many variant small molecules with the strained four-membered β-lactam or some other strained ring could have been an effective antibiotic, but it is more likely that the original biosynthetic pathway made only one kind of molecule. Subsequent events led to the modification and diversification of that species into different four-membered β-lactams (i.e., assorted penicillins and cephalosporins) so that different molecular species with the same β-lactam grouping could be more effective under various circumstances. Of course, phar-

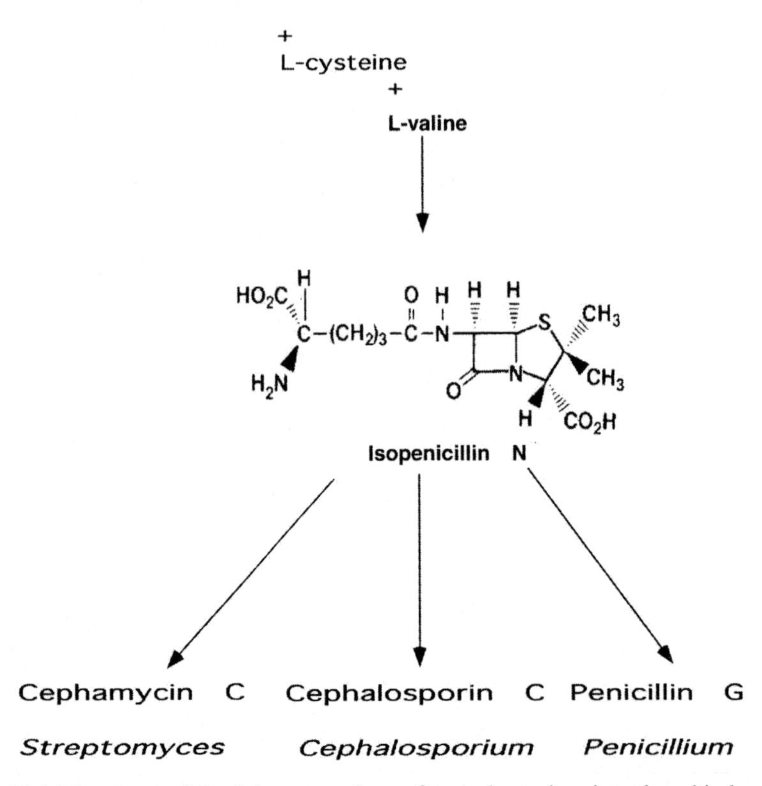

Figure 19.3 Movement of the β-lactam pathway from a bacterium into three kinds of fungi. Once the pathway entered into these fungal classes, many minor variations developed.

maceutical chemists created even more β-lactams, leading to the evolution of variant β-lactamases by bacteria.

β-LACTAMASES

It has been thought that only fungal species could produce penicillin, because *penicillium* species could safely produce penicillin as their cell wall makes them naturally immune. However, it would be lethal to bacteria. Still there are β-lactam-producing bacteria and the evidence indicates that the pathway developed in a streptomycete. So if bacteria produce a β-lactam, how do they avoid its toxic effects? A plausible explanation is that both the β-lactam for export and a β-lactamase for self-protection could have co-evolved in the first successful antibiotic-producing bacterium. It could then have a low enough level of internal

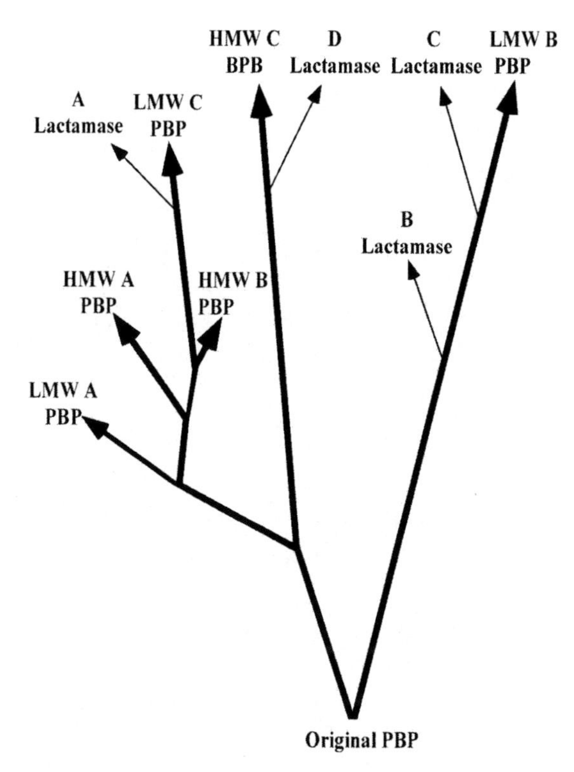

Figure 19.4 The evolutionary tree of penicillin binding proteins (PBPs) and β-lactamases. Taken from Koch (2000).

antibiotic, but still excrete enough antibiotic to inhibit or kill the surrounding bacteria.

It is generally the case that a toxin-producing cell has, and in fact, must have, ways to protect itself against its own toxin. All streptomycetes, of course, do have PBPs that are inhibited by β-lactams; however some also produce β-lactams. We can imagine that some streptomyctes, possibly more than a billion years ago, developed an enzyme system to manufacture β-lactams, and also developed a β-lactamase via duplication and variation of a gene encoding a PBP. This simple generation of β-lactamases apparently only occurred several times in the history of life on this planet (Figure 19.4). We will see below, however, that this is not too difficult an evolutionary task. From it a variety of relatively minor variations developed, some early on and numerous others now in the antibiotic era (see Bush, 1997).

I suggest together with others that when a β-lactam antibiotic was first produced it did not serve to protect a target cell against an antibiotic produced by an attacking organism, but was produced instead within the attacking organism.

This would serve the producing organism's needs to avoid toxic internal levels of the antibiotic being produced for export. Later on, lateral transfer brought antibiotic production to the target organisms. Then subsequently, by theme and variation, a diversity of β-lactams were developed so that now there is a panoply of both β-lactams and cognate β-lactamases.

The bacteria targeted by secreted β-lactams were under strong selective pressure and the development of resistance mechanisms became essential. The DNA of many penicillin-binding proteins (PBPs) and related molecules has been sequenced, and from the phylogenetic tree (Figure 19.4) that developed from these studies, it can be concluded that the development of β-lactamases occurred independently several times.

RESISTANCE MECHANISMS IN GENERAL; DEVELOPMENT OF β-LACTAMASES IN PARTICULAR

There is something special about β-lactam antibiotics that do not apply to most other biologically generated toxic substances. The evolution of a resistance mechanism must have been by a very difficult and a long step-by-step process. Long times would have been involved because a series of mutations is required to produce the very complex resistance mechanisms against naturally produced antibiotic substances.

It is at first surprising that few clinically or agriculturally relevant antibiotic resistance mechanisms appear to be newly developed (Koch, 2000). Indeed, for most antibiotics that have been used by humans, the resistance mechanisms are not new, but have pre-existed somewhere in the world; the effect of antibiotic use by humans was to lead to lateral transfer of resistance genes to pathogenic and agriculturally relevant microorganisms. In the important cases of the β-lactamases and MRSA (see below), the mutations are certainly old.

The β-lactamase proteins are special amongst agents for antibiotic resistance because there is very little biochemical difficulty in slightly modifying a PBP gene that codes for an endotranspeptidase and changing its chemistry so that it now encodes a hydrolyase (Koch, 2000). These transpeptidases form tail-to-tail bonds by exchanging the terminal D-Alanine of D-Alanyl-D-Alanine from a penta-muropeptide for an amine present in another muropeptide (in the case of *E. coli* this amino group is part of the zwitter ion part of diaminopimelic acid). It is important to emphasize that this arrangement can occur because each "transpeptidation" event is powered by the prior incorporation of the terminal D-Alanine. This is important for the energetics of the reaction, its reversibility, and the enzymatic (i.e., hydrolytic) interaction with water.

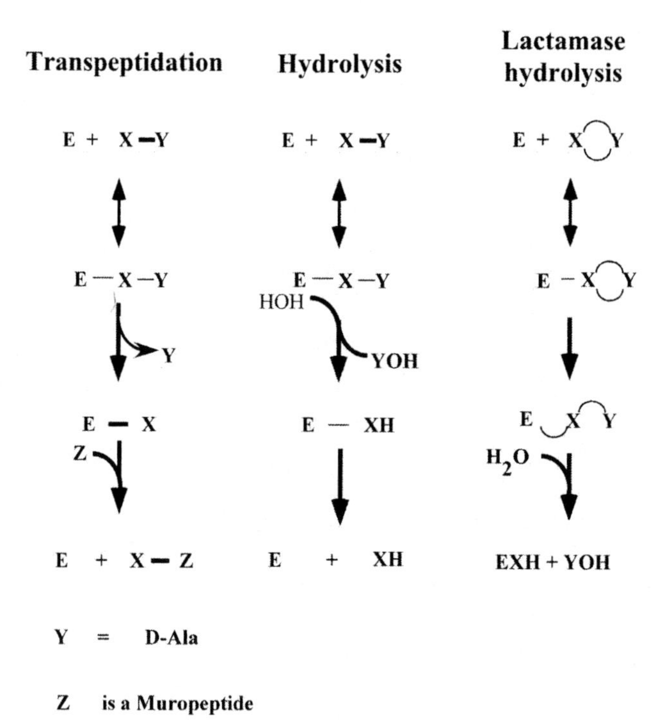

Figure 19.5 Enzymatic pathways of three types of enzymes that deal with a chemical bond that may be called in particular circumstances an amide, a peptide, or a lactam. However, these are different and lead to different outcomes.

Only a few point mutations suffice to let water into the active site so that, instead of the D-Alanyl-D-Alanine reacting uniquely with its cognate special amine, it could react with the more ubiquitous H_2O. A few specific structural changes would allow water to be used instead, and thus a few simple changes convert the enzyme from a transpeptidase into a protease (hydrolyase) (Koch, 2000). Transpeptidases (transamidases) act by binding a substrate, cleaving the critical bond and reforming the linkage, but with a new partner of equivalent energy. In the case of bacterial cell wall enlargement, the D-Ala-D-Ala bond is cleaved and the tetra-muropeptide remains bound to the enzyme until another muropeptide supplies a replacement amino group, which for the case typified in Figure 19.5 is the free amine end of a diaminopimelic acid residue. The β-lactams instead use water.

Since hydrolyases use H_2O instead of an amino group as an acceptor, it is possible that rare errors could be made by the endotranspeptidases so that water was used by mistake. This would still cause D-Alanine liberation by hydrolysis. This aberrant proteolysis would prevent local cross-bridge formation, but would only prevent formation in a small part of the wall. On the other hand, hydrolysis of the four-membered strained ring of a β-lactam by the β-lactamase enzyme destroys the antibiotic and would occur repeatedly at a high rate. A few molecules of β-lactamase can thus destroy many drug molecules and could protect the target cell. These β-lactamases would not be active on the tail-to-tail bonds of murein but could cleave any β-lactams that binds them. Note that β-lactamases must not function as autolysins and lyse the murein, so their specificity may be dependent on the strained state of the stressed ring because the bond split in the lactam is an amide bond, just as in a peptide bond.

This is an important idea for the future of clinical microbiology and it will be stated a little differently: Because β-lactamase action is similar (except for the acceptor) to the transpeptidase action that is needed for saccular expansion, PBP for growth must have been duplicated. Then at several points in the evolutionary tree, a PBP turned from an endotranspeptidase into a β-lactamase by changes that allowed H_2O into the active site. Such enzymes are also known as D,D-hydrolyases or D,D-carboxypeptidases. Considering how easy this is theoretically, it is not surprising that it happened quite early, and in variants in many lines of descent. The available amino acid phylogeny provides evidence that all the β-lactamases came from one PBP (Koch, 2000; Massova and Mobashery, 1997; Massova and Mobashery, 1999; Medeiros, 1997; Barlow and Hall, 2002) and that they did not arise multiple times. The amino acid sequences encoded by the genes in this large family of penicillin-recognizing enzymes have several highly conserved sequences. This group includes the PBPs per se, penicillin-sensing proteins, and β-lactamases. This has led to the development of evolutionary trees (Bush and Mobashery, 1988a, 1988b). All these proteins together form a superfamily (Figure 19.4). The point relevant to the evolution of resistance, however, is that the β-lactamases branch later than most PBP families. This clearly suggests that these β-lactamases did not exist until some time after murein walls were ubiquitous and only, presumably, after penicillin-type antibiotics became common. It can be concluded that β-lactamases were common in the bacterial world millennia before the time of Fleming. Quite recently this has also been found to be the case for a different group of β-lactamases, the OXA group (Barlow and Hall, 2002).

LACTAMASE-RESISTANT LACTAMS EXTANT BEFORE MEDICINAL CHEMISTRY

Before the antibiotic era of medicine there was further evolution in various lactam-producing organisms to create novel and variant compounds. This probably occurred in response to the resistance arising in the target organisms.

These variants still had the four-membered lactam ring characteristic of penicillin and are classified as penicillin or cephalosporin-type compounds. There appear to be a large variety of compounds based on the four-membered ring, varying in the side groups, each presumably has different secondary effects. It is also clear that natural antibiotic production entered only a few lines of fungi and within these lines many variant β-lactams subsequently developed. This is consistent with the idea that these variations were driven in response to the development of β-lactamases in the target organisms.

OVERVIEW OF EARLY β-LACTAM EVOLUTION

Competition for space and resources and the potential that bacterial organisms could serve as resources for other organisms made the development of β-lactams inevitable after the development of the peptidoglycan sacculus and the consequent success of bacteria. Ever since it arose, the existence of a sacculus is the bacteria's strength and also its weakness. The critical bond tail-to-tail is shown in Figure 19.6. The bacterial sacculus became the corner stone of the bacterial survival strategy because of the achievement of the formation of a strong wall outside the basic cell through the proper application of mechanical and chemical engineering principles. To accomplish this feat, free energy resources had to be built into the precursor units inside the cell before export to the outside of the cytoplasmic membrane where the sacculus was being enlarged.

In the previous section it was noted that a class of pre-strained, reactive, PBP-binding inhibitors of the PBPs; i.e., the β-lactams, were and had to be antibiotics and could irreversibly inactivate the PBPs and prevent bacterial growth. In response the bacteria would develop responses.

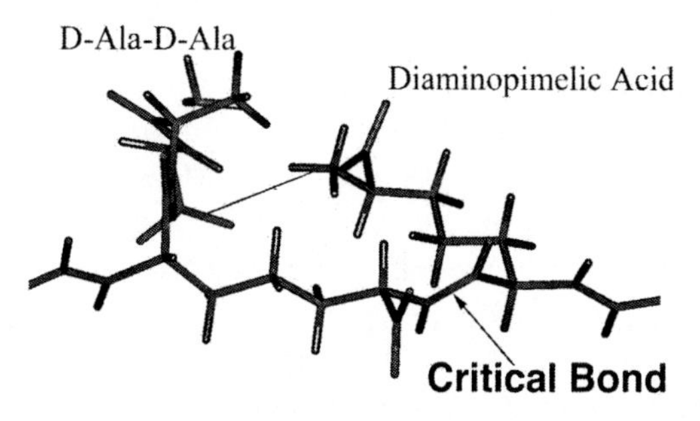

Figure 19.6 The critical bond formed by the endopeptidase PBPs that are the target of β-lactams. This linkage is marked with an arrow). Also shown is the relationship of the amino group of the diaminopimelic acid and the carboxyl group of the D-Alanine. This charge interaction occludes entry to the critical linkage area until stress is applied to the nonamuropeptide.

One counter-countermeasure strategy is as follows: when the cell is subject to attack of its essential PBPs, the attack could be overcome if the cell had an alternative molecule similar to the usual PBP, but with special properties. It would be similar to the PBPs needed for cell wall growth, but would have a poorer ability to conserve chemical energy. It would, therefore, be a hydrolyase and not a transpeptidase. The important point is that the β-lactamase could be recycled and would react irreversibly with the penicillin derivative and protect the endotranspeptidase PBP. The β-lactams and β-lactamases interacted with each other in evolutionary history, each prompting the other's modification and diversification over geological time.

TURGOR PRESSURE RESPONSES TO WALL ANTIBIOTICS

It is obvious that turgor pressure inside cells is important for many aspects of bacterial physiology and for the action of wall antibiotics; however, methods to measure turgor pressure are poor. A light scattering apparatus and method were developed and used to study turgor pressure and the effect of wall antibiotics. Only a short summary can be given here; Koch and Pinette (1988) give a more detailed review.

The developed analytical technique can only be applied to freshwater Gram-negative organisms that have the useful property of having gas-filled vesicles. No other known or commonly studied organisms were appropriate for this specialized technique. On exposure to adequate pressure the gas vesicles collapse, the gas rapidly dissolving in the cell contents, and the light scattered by the cell is greatly reduced. The apparatus constructed exposed the cells to a ramp of pressure and recorded the light scattered at right angles to an incident laser beam. The experiments exposed the cells to various agents and measured the pressure that collapsed one-half of the vesicles.

Before the experiments were done, the expected results were almost trivial. The assumption is that although the antibiotic would prevent wall enlargement, other cellular processes should continue, such as uptake of solutes from the medium and macromolecular synthesis. This unbalancing of cell processes should increase turgor pressure. Antibiotic treatment should thus decrease the amount of added pressure from the experimental apparatus needed to crush the vesicles.

Under ampicillin action, the cell is expected to rupture and the turgor pressure differential should abruptly become nil. Of course, a larger hydrostatic pressure would have to be supplied by the apparatus to crush the gas vesicles. As can be seen from Figure 19.7, this is not at all what happened, at least not to all of the cells. The collapse curve occurs 20 minutes after antibiotic treatment changed

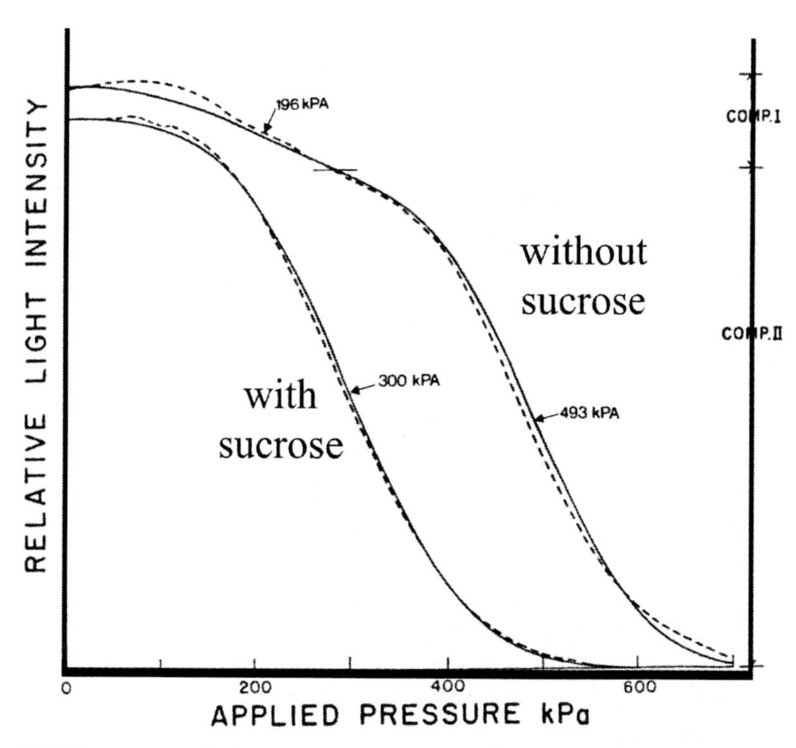

Figure 19.7 Two classes of cell pressure responses in cells treated with and without ampicillin. *Ancylobacter aquaticus* cells were examined in the described light scattering apparatus. Without ampicillin treatment initially the vesicle collapse curve was monophasic like that marked "with sucrose" but required a higher pressure for collapse. After 20 minutes of treatment with ampicillin, the curve changed to the bi-phasic one marked "without sucrose".

from a monophasic to a biphasic shape. From the pressure at the midpoint of the two components, it could be concluded that some bacteria had done what was expected and their vesicles now had a collapse pressure similar to that found in the presence of a large concentration of sucrose where the cell's turgor pressure would be zero. It followed, therefore, that the cell envelope must have ruptured. Surprisingly, the collapse curve of the remaining cells had a stable turgor pressure, in fact, one a little higher than the untreated cells. This has to mean that in the face of a challenge that prevented the enlargement of the stress-bearing wall, this subset of cells had almost stopped the further increase in turgor pressure. To prevent a blowout, the cells must have blocked many active transport systems or opened leaks to let ions and solutes return to the medium. They may also have had to shut down synthesis of macromolecular species.

If this type of mechanism is a general trait of bacteria, including pathogens, and is applicable to the many antibiotic agents that prevent wall growth, then this is quite a different problem than has been previously recognized and one that will need to be addressed by the biomedical community in the future.

MRSA (METHICILLIN-RESISTANT STAPHYLOCOCCUS AUREUS)

The other case of an apparently new evolution of an antibiotic is the emergence of methicillin resistant *Staphylococcus aureus* (MRSA). Although MRSA has been very important clinically since 1961, there is reason to believe that the basic *mecA* gene, encoding PBP 2′; evolved from an existing PBP in the distant past as a supplement to the normal group of PBP.

It is reasonable to imagine that PBP 2′ would be less susceptible to inactivation by β-lactams and protect against natural lactam antibiosis when the inhibition of wall synthesis by lactams from the environment is modest. When present in most (non-clinical) organisms, it serves to supplement the usual PBPs and not to replace them. Selection for function in these cases occurs when the cells are subject to mild blockade by β-lactams. PBP′ 2 probably is not fully efficient in wall synthesis, and therefore its presence is an extra operating cost to the cell, but not a severe one.

In the pre-antibiotic era, PBP 2′ was not maximally expressed because there was repressor control over its synthesis. This repression resulted from the action in the operon of *mecI* and *mecR1* that included it. These genes moderate the production of PBP 2′, but presumably allow enough to be synthesized in the presence of low levels of β-lactams that was produced by nearby organisms. However, once methicillin and oxacillin were commercially produced and their medical use expanded, the partially repressive conditions that had led to its evolution did not permit high enough levels of PBP 2′ for effective resistance. As a result, the regulatory genes have been lost and the genetic material fused to a lactamase gene cluster with appropriate regulatory capability. Such changes can easily and rapidly occur.

The reason for believing that the basic evolution of MRSA occurred prior to 1960 is that *mecA* is found in other organisms in a gene cluster together with its regulatory genes. This implies a long evolutionary time was needed to create the gene and the controls for its action in the pre-antibiotic era. The critical point is that *mecA* did not evolve in the last 60 years; rather, the regulatory genes were lost by inactivation or deletion and allowed resistance to stronger challenges to be mounted shortly after the beginning of the antibiotic era and the usage of methicillin and oxacillin.

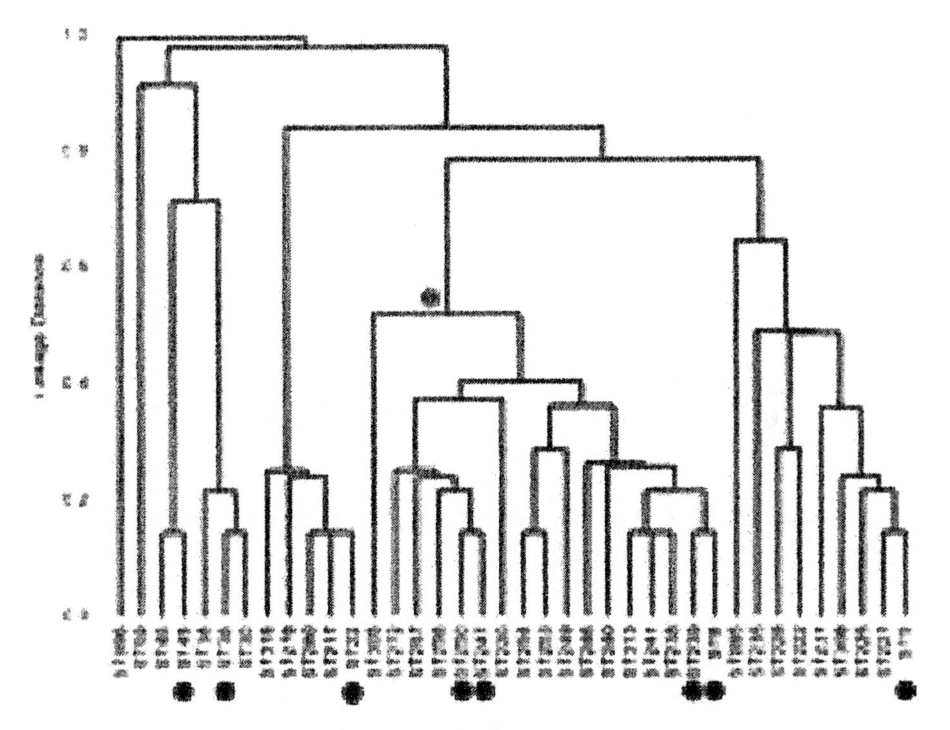

Figure 19.8 The phylogenetic tree of methicillin resistant *Staphylococcus aureus* MRSA strains. This graph is reprinted from Enright *et al.* (14). It shows the divergence of modern clinical strains from the original MRSA. The original one is shown at the left- hand side of the figure. At least eight major strains evolved independently and then diverged but retained this MRSA character.

This is not to say that there have not been many changes in MRSA in our era. Spratt and coworkers (14) analyzed 359 MRSA strains and separated them into 38 families. These changes all happened recently, but presumably the parent of all, the original *mecA*, goes very far back in evolutionary time. This pedigree of existing organisms is shown in Figure 19.8.

LACTAM ANTIBIOTICS TODAY

β-LACTAMASES DESTROYING β-LACTAMS IN THE CURRENT ANTIBIOTIC ERA

No one could foresee the consequences, both great advances and failures, that we now can see looking back. Literature on the development of penicillin is fascinating. The conclusion to be drawn is that this is a case where history repeated itself many times and may repeat itself much more forcibly in the time to come. The defeats came from genes existing in the world biosphere, lateral transfer that mobilized these genes, and mutations that modified the antibiotic action of pathogens. The action of extant genes transferred from host organisms remote from human pathogenesis defeated many potential clinical agents and is defeating specific antibiotic actions now. This kind of event is likely to happen again and again. How to avoid this in the future is not obvious, but there are certain design features that may be important.

INITIAL MANUFACTURE OF PENICILLINS AND THE MICROBIOLOGICAL SEQUELAE

When penicillin G was first used to treat infected people, it was so valuable and amounts were so limited that it was recovered from the urine and reused. Penicillin V was developed; it was more acid resistant and could be taken orally. Quickly however, resistance became apparent. The first observation was the finding by Abraham and Chain that some strains of *Escherichia coli* were already resistant, and shortly after that, Kirby (28) found resistant strains of *Staphylococcus aureus*. These resistance mechanisms, in hindsight, were no doubt generated by the prior existence of β-lactamases present in various microbial populations. Of course then, and especially now, the vast majority of the staphylococci had, and especially now continue to have, β-lactamases borne on plasmids. It is important to note from the discussion above that the creation of MRSA was a man-made event. This assertion can be made because streptococci that do not have β-lactamases were treated with penicillin. On the other hand staphylococci, which originally only rarely had β-lactamases but coexisted in the same environment, must have acquired the lactamases because of the lactam treatment. The consequence was that quickly and commonly staphylococci have acquired the MRSA gene (59).

While the β-lactamases are all generally related, indicating that they all go back to the primordial PBP, there are many variants and their sequences suggest that a large number of β-lactamase genes in the world's biosphere existed prior to human involvement in chemotherapy (51, 52). Bush and Mobashery (5) counted 255 primary sequences and it is clear that many of them arose before our era and took time to be transferred to medically significant bacteria. Of course, variations arose or were subject to minor modifications after the human exploitation of β-lactams. This is also the known history of the OXA β-lactamases (2) deduced from available sequences.

NOVEL β-LACTAMS

In the late 1950s, several new β-lactams were discovered, such as cephalosporin C and 6-aminopenicillianic acid. These permitted chemists to produce lactams that resisted some of the lactamases. They included the semi-synthetic isoxazoly penicillins such as methicillin, oxacillin, and the "first generation" cephalosporins. The use of cepalothin, cephaloridine, and cefazolin significantly decreased the role of β-lactamases that was then rampant in Gram-positive organisms.

EXTENDED-SPECTRUM BETA LACTAMASES (ESBL) AND CEPHALOSPORINASES

Not surprising in hindsight, antibiotics including methicillin and oxacillin were associated with the appearance of extended-spectrum β-lactamases (ESBL) and cephalosporinases. In the middle of the 1960s, new plasmid-encoded β-lactamases appeared in Gram-negative organisms that had been subject to cephalosporins. These enzymes have been designated SHV-1, TEM-1, and TEM-2. Of these, TEM-1 has caused particularly grave medical problems. This danger led to a concerted attack on TEM-1 by the pharmaceutical industry and resulted in a two-pronged approach. One approach was the development of β-lactams containing clavulanic acid and penicillin-derived sulfones. These were offered to patients together with a β-lactam that blocked certain PBPs. This worked because the clavulanic acid irreversibly tied up the β-lactamase, thus protecting the β-lactam antibiotic, which was then able to inhibit PBPs and block wall growth. The alternative prong was to find and use new lactams that were resistant to the β-lactamases like the carbapenems. Later, some of the other "third generation" cephalosphorins and still later the monobactam, aztreonam, were developed.

WHAT ABOUT THE FUTURE

The development of a new genetic capability can be rapid or extremely slow; it can be a common and an often-repeated change or it can be a unique occurrence. Consider these unique rare events: the origin of life, the origin of bacteria, the origin of β-lactams, and the origin of the original MRSA mutation. These were events that apparently happened only once in the history of the world. Once these "saltations" or "macromutations" took place, then propagation, expansion, and variation occurred. [A variety of more ordinary mutations could occur more easily once a major advance had taken place and often tune variants to create distinct organisms or strains.] AU: Is there something missing in the last sentence? This, in general, led to the generation of improved measures, but also to countermeasures to these improvements. All these smaller changes usually are only relevant in the context of the originating "macromutation". Although usually not "a great leap forward", the subsequent changes are usually minor, but still offer a selective advantage. Figure 19.8 shows an example of this. One ancient mutation in *Staphylococcus aureus* made it resistant to β-lactamase with the creation of the original *mecA* mutation, but then further evolution in various target strains as well as in the antibiotic-producing organisms made further changes necessary and readily possible.

Once a β-lactamase had originated from existing PBPs, it was evolutionarily easy to develop many more slightly different β-lactamases and once the first MRSA had arisen, it was relatively easy to develop MRSA variants. From these two revolutionary ancient beginnings, these macromutations also made the bacterial responses to man-made agents evolutionarily easy. It led to the possibility of artificial variation of lactams using genetic methods, natural products chemistry, semi-synthetic chemistry, and synthetic chemistry. We have discovered that it is hard to find long functioning counter-countermeasures against these bacterial responses that are so well established and have functioned for such a long time.

Chapter 20

Antibiotics and Resistance, with an Emphasis on Aminoglycosides

The purpose of this chapter is three-fold, (1) to present the general situation for antibiotic production and resistance, (2) to focus on biology of aminoglycoside antibiotics that block protein synthesis, and (3) to suggest strategies to enhance the effectiveness of this class of antibiotics so that they can have a longer useful life within the clinic—even when they are heavily used.

ANTIBIOTIC RESISTANCE MECHANISMS

It is necessary to distinguish between destructive chemicals, antiseptics, and antibiotics. An agent that kills organisms non-selectively, such as phenols, alcohols, cationic antiseptics, and heavy metals, can be useful if their action can be directed by choices of where and how it is applied. An antiseptic is only a little different because of the topical way it may be applied. An antibiotic must have the property that it acts against the microbe, and at least much less against the human or animal that is also simultaneously exposed to it.

Obviously useful antibiotics should not act against mammals, such as humans. This means that they must operate on a difference between bacterial and eukaryotic metabolism. This is why the bacterial murein is such a preferred target, because it is unique to bacteria. However, nucleic acid synthesis, protein synthesis, and action against membranes are other targets where a difference does exist. These are useful targets because, although in the three domains there are great similarities between them, there are sufficient differences so that agents that distinguish between the differences do exist. The next chapter deals with future possibilities with old and future β-lactams and murein as the target, but here we deal with one group of inhibitors that acts more on bacterial protein synthesis. Moreover, for this one group we will focus on three different time frames taken from the story of aminoglycosides.

ANTIBIOTIC RESISTANCE EVOLUTION WITH AND WITHOUT MAN'S INTERFERENCE

When life first arose, there would have been no purpose for antibiotics and the biosynthetic pathways to make them would not have developed. It has been argued (Koch, 1993) that up to the time of the Last Universal Ancestor (LUA), there would have been no stable diversity, and life was essentially a monoculture. Carl Woese feels that at this time there was much diversity, even though it was not stable. Self-evidently, from the phylogenetic tree there was no stable diversity below the root of the tree. With the establishment of domains and stable diversity it was possible for apparently competitive organisms to co-exist because the competition was not exact. With diversity, niches developed for organisms of a variety of properties, sizes, and strategies. Predator, prey, commensal, symbiotic, and antibiotic-producing organisms now all could prosper. The establishment of niches (or roles for different organisms) and habitats (or different locations and environments for different organisms) was exploited and is necessary for the development of complexity in an ecosystem. Antibiotic compounds would then have had a functional role and, no doubt, a range of organisms produced very large numbers of antibiotic chemicals. Since intermediary metabolism develops by theme and variation of existing enzymes, families of related natural products would have arisen.

With the introduction of natural antibiotics, selection for resistance mechanisms occurred in the target organisms. Mechanisms of different types developed in a number of broad classes. The major categories are: reduced permeability, inactivation of the antibiotic, alteration of the target within the host, development of a substitute (but resistant) biochemical pathway, and development of efflux pathways. The different types of resistance mechanism have been considered in many reviews. I will single out my own review of 1981 for two reasons. First, it listed all the classes now known, except multidrug resistance, which was not known at that time. (The mechanism for the efflux of tetracycline was known, but the existence of mechanisms that would cause the efflux of a variety of quite different agents was not yet contemplated or hinted by our then observations or experiments.) This success of science, as it had developed twenty years ago to have found all the varieties of classes now known except that of the general efflux pathway mechanisms, suggests that maybe there are no unrecognized classes of resistance mechanism left in the biosphere. This is parallel to the general finding that search of the biosphere for a new antibiotic (of known classes) has a low chance of success nowadays. Currently new molecular biological techniques to scan organism's genomes and genes directly are being used to search further for antibiotics using high throughput techniques. Besides members of the known classes and the targets of known antibiotics, they may find novel classes, however, they may not. Here the alternative is suggested that it may be that man may be able to devise and synthesize new types of antibiotics directed towards targets for which there are no natural antibiotics and, therefore, no natural resistance mechanisms present in the biosphere.

SLOW, LABORIOUS EVOLUTION OF NEW *DE NOVO* PROTEINS

Much of *de novo* evolution has occurred by duplication of a gene and then modification of one copy for a new or an improved purpose. Evolution by a series of point mutations has limitations. If a new purpose is quite disparate from the genes original one, then further tuning and molding of the protein structure is needed. It must be assumed that development proceeds slowly and that duplication, alternate activation and silencing occurred repeatedly. While duplication and selection were the only evolutionary procedure initially, when mechanisms for lateral transfer arose (within cells and between cells) rearrangement of genetic regions and fusing with still other genes must have soon become as important as they are today. In several papers, I have considered the kinetic effects on enzyme evolution of: reversibly inactivating a gene (Koch, 1972), periodic selection (Koch, 1974), and possession of a tandem array of genes (Koch, 1980). With such movement of DNA from cell-to-cell mediation of this transfer both for the aid of the host and the pathogen, the net effect would have been the inhibition of *de novo* evolution of the host's resistance mechanisms because lateral gene transfer would dominate.

There is another quite relevant point that needs to be stressed: it appears (Koch, 1972, 1996) that the natural evolutionary mechanisms on this planet never approached the most efficient possible protein structures. It did not scan all the possibilities. This failure is demonstrated by the contemporary results with modern recombinational techniques. With these modern laboratory genetic techniques, random replacement of the codons in specific parts of a gene can be used to find much more efficient proteins than exist now in the biosphere. These combinations had just never been "tried" by ordinary evolutionary mechanisms lacking human intelligent input and methods. The biological implication of the results of the "selex" experiments of the last decade is that Nature was not able to explore sequence space anywhere near as completely as the modern biochemist can because he or she needs only to scan relevant local regions and not the whole gene.

LACK OF EVOLUTION OF NOVEL ANTIBIOTIC RESISTANCE AFTER MASSIVE HUMAN USE OF ANTIBIOTICS

The second point exhumed from this old review is the assertion that no radically new (*de novo*) resistance mechanisms had developed in the then 40 years of the antibiotic era of medicine. In fact, the resistance mechanisms important in 1981 had been occasionally generated by simple genetic changes, but mainly had pre-existed in the antibiotic era and represented mainly lateral gene transfers to pathogens from other organisms. Source organisms would include: the producing organisms of antibiotics (as in the case of aminoglycosides, discussed below) and from those organisms whose enzymatic system were inherently resistant. Considering the massive nature of the application of antibiotics and the strong selection pressures during the last half century, it appears at first strange that extensive *de novo* evolution did not appear. I have argued (Koch, 1981) then and emphasize here that this lack of evolution was because the gene transfer from other parts of the biosphere preempted *de novo* mutations that otherwise would have improved the basic resistance mechanisms.

ROLE OF TRANSFORMATION, PLASMIDS AND VIRUSES

It has been argued that insertion of DNA into living cells is not trivial and that there are no passive, spontaneous ways for DNA to enter a living cell (Koch, 1994). Although a half a dozen quite different mechanisms for gene transfer exist in modern organisms they are all quite elaborate and complex. The transfer mechanisms need to be complex because a highly polar nucleic acid has to be passed through a hydrophobic lipid bilayer. In the modern world, transmissible plasmids and viruses can do it, some bacteria under certain conditions do have genetic mechanisms allowing them to take up naked DNA, and of course, a sperm can pass its DNA to the accepting egg by a very sophisticated process. So the argument is that before the development of these gene-transfer mechanisms, much *de novo* creation and perfection had to take place to evolve antibiotic resistance mechanisms inside a particular cell and such ability was not passed laterally from cell to cell until much later.

After lateral transport evolved and could take place efficiently, then it could happen that some possibly poorly effective, but existing mechanism, from almost anyplace, could out compete the *de novo* development or improvement of an existing mechanism. This would impede *de novo* evolution generally. Although production of antibiotics in nature probably was never near the intensity of industrial production and clinical use, mobilization of a resistance gene to a targeted microorganism by lateral transport of antibiotic resistance genes would have occurred in the past. However, such movement was not at the breakneck pace of the last 60 years.

SMALL BUT IMPORTANT CHANGES IN THE CURRENT ANTIBIOTIC ERA

It has been suggested at many points in this book, in Koch (2000), and it is obvious that the development of an antibiotic led to the development of resistance countermeasures. The antibiotics led in the distant past to abstraction and variation, for example, of the details used by producing organisms for antibiotic resistances. Cephalosporins and penicillins differ in small ways such that their evolution is a small development compared with the development of the basic β-lactam ring. The latter was a major and unique event in the bacterial lineage. The pharmaceutical chemist by mixing, matching, seizing available compounds from a range of organisms, and using several compounds in combination, has increased the stress on bacteria in certain habitats. The organisms there, in turn, have acquired enzymatic mechanisms from the biosphere and have mutated the acquired genes to become more resistant. These steps represent very simple evolution compared with the biological processes that had occurred and preceded the application of antibiotics for clinical and agricultural use.

THE EARLIEST RESISTANT MECHANISM, A VERY SPECIAL CASE

This chapter may appear as an aside in a book whose particular attention has been to murein metabolism, but I hope to point out some general principles of the development of antibiotic resistance mechanisms. In this section, first, I will consider the case of streptomycin resistance (the earliest found aminoglycoside) as a paradigm for the action, and mechanism of natural antibiotics.

The first studied resistance results from a very rare mutation that alters the protein S12 of the ribosome. This site of action was detected two decades later than the Waxman discovery of this antibiotic. Discovery of the genetics of this resistance, by the way, was the first demonstration of a genetic mutation in bacteria. This was the work of Demerec (1948) using the Luria and Delbrück fluctuation method (1943). Although in the 1940s and 1950s, genes and their mutability had been accepted as the basis of living matter in plants and animals for 50 years, it was only in the 1960s that there was the full acceptance of a genetic basis for the life of microbes.

The S12 mutation makes protein synthesis considerably slower, but much less error prone (Dykhusien, personal communication). In the absence of streptomycin, when a competition experiment between the wild type and the streptomycin-resistant mutant is carried out, the fast, but error prone, wildtype rapidly replaces the slower growing mutant form, even if the ribosomes containing the S12 mutant make fewer mistakes during protein synthesis. This suggests that an occasional error in protein production is not as detrimental for growth as is a slow speed of protein production.

The development of a resistance mechanism by mutation of the S12 protein is of little clinical importance today. This is simply because the plasmid-borne detoxification mechanism does move rapidly from unknown sites in the biosphere to protect a pathogen residing in an antibiotic-treated host. Because the S12 mutant, although resistant to streptomycin, grows slowly in the presence or absence of streptomycin, it is cleared by non-specific resistance growth-selection mechanisms of the host while the plasmid-born detoxification mechanism is not selected against in growth rate competition with the sensitive organism.

It is not known which one—the fast and imprecise or the precise, but slowly, growing version—was the original or primeval gene. Did the original wild-type organism exist before it became a target of streptomycin? However, it is clear that in the isolated laboratory culture in the presence of streptomycin and isolated from the rest of the world, many populations would be large enough to have the rare mutant at the time of challenge, and therefore, it will develop to become a streptomycin-resistant culture. The mutants will take over the decimated population and continue to grow. Later, when the antibiotic challenge is ended, then slowly, but eventually, the rapidly growing sensitive wildtype cells will replace the S12-mutants cells. (Presumably, this is the common case that is a solution to a novel specific problem and degrades the rate of growth because it is deleterious in some other way.)

The point of this subsection is to point out a number of important aspects of the microorganism's response to antibiotics in the environment. It is likely that there are a range of streptomyctes that have to be immune to the action of the antibiotic they produce. This is a subject to which we will return in the next subsection. Although the evolution of a chromosomal streptomycin resistant version of S12 may well have occurred in producing organisms, it is difficult to imagine how, before there was a lateral gene transfer mechanism, the resistant gene could get to the target organism. If it did, it must have been an extremely rare event. More likely there were only *de novo* mutations at the time so only a chromosomal resistance mechanism that could serve this purpose could have arisen in the sensitive organisms. Although the change to resistance protected during challenge and slowed bacterial growth in the absence of challenge, this is not a grave problem to the target bacteria, because the wildtype will re-arise. This implies that the S12-mutant state was not an earlier form but only developed after many cycles of production and loss or partial loss. Consequently, the subsequently evolved mode that we now see has sped protein synthesis, was at the expense of accuracy and is relatively easy to change to the streptomycin-resistant accurate form. The streptomycin-resistant form is more precise and slower. Obviously, speed is more important than precision, in this example. The chromosomally resistant form is due to a rare mutation which can easily be replaced by being outgrown by the faster and less precise wildtype, while the (clinically important today) plasmid-born mechanism that destroys the antibiotic is both easily found and easily lost.

THE LATER CONTEMPORARY HISTORY OF AMINOGLYCOSIDES

From 1944 until 1970 a series of aminoglycosides (streptomycin, first) were found largely in *Streptomyces* and also semisynthetic derivatives were developed. But as mentioned, resistance due to chromosomal mutations were not clinically an effective resistant-mechanism for most diseases. Instead, plasmid-based enzymes that detoxified the antibiotics were the ones that became important. These enzymes fell into three classes: O-nucleotidyltransferases, N-acetyltransferases, and O-phosphoryltransferases. The O-nucleotidyltransferases typically add AMP from ATP. The N-acetyltransferases add Acetyl-CoA to make N-acetyl-aminoglycosides. The O-phosphoryltransferases use ATP to phosphorylate the antibiotics. There are many versions of each of the three types so one should conclude that there were many independent creations of these kinds of detoxifying enzymes. It is clear that most, or at least many, of these enzymes came from streptomycetes bacteria that produced aminoglycosides (Beveniste and Davies, 1973). All of this history is consistent with the idea that as aminoglycoside-producing organisms arose and diversified, each strain had to have enzymes for its own self-protection. This in turn led the target organisms to acquire the enzymes and may well have been a force leading to development of transmissible plasmids to mediate the transfer.

Chapter 21

Future Chemotherapy Aimed at the Bacterial Murein

The purpose of this last chapter is two-fold, both to suggest strategies to look for antibiotics that should have a long useful life even when heavily used and to focus on the metabolism of bacterial walls as a still rich area to search and find useful antibiotics. Parts of the wall growth process are not now targets of known natural or man-made chemotherapeutic agents but could be susceptible to new chemical agents. Although such agents may be hard to find, once they are found they will be less susceptible to losing their effectiveness through the development of bacterial resistances. This thought is based on the presumption that if such types of antibiotics do not exist now, then resistant mechanisms against such a class of antibiotics would not exist anywhere in the world, and consequently a counter-measure would not exist at all and would not be available to be acquired by pathogens through lateral gene transfer. If such a hoped-for new agent could be created in the chemical laboratory, specific resistance to it by the community of microorganisms would require de novo evolution. This, it is assumed, would be a time consuming process for the bacterial community.

It could be that the targets for new antibiotics are in an area of metabolism that is not well understood. Especially, this is the process of transporting the unit disaccharide penta-muropeptide through and then integrating it into the growing wall.

If such hypothetical antibiotics were realized in the future they would have a medical action that would be without parallel. They would be expected to serve man's needs for a long time and not quickly lose their usefulness. Of course, the role of non-specific resistant strategies, such as the action of multidrug pumping systems that could act non-specifically, would have to be anticipated, considered, and avoided.

ANTIBIOTIC RESISTANCE MECHANISMS

Before transmissible plasmids and viruses arose and at a time when transmission of a resistance gene from organism to organism was nearly or totally impossible, resistance could only occur by mutation of the target genes of cells. A number of such chromosomal resistance mutations are known today: resistance to rifampin by a change in the RNA polymerase; resistance to erythromycin, coumermycin, and streptomycin by modifying ribosomal proteins; and resistance to the inhibition of DNA gyrase by quinolones. These are all changes of the cellular target, but in general this strategy is a poor alternative relative to lateral migration of a gene from cell to cell when both modes are possible. Davies (1997) gives a list of resistance mechanisms to antibiotics classified by chromosomal mutation and/or by gene acquisition. Numerically, there are a few more of the latter types than of the former. The biochemistry for chromosomal mechanisms can be categorized, largely according to the possibilities pointed out by Davis 40 years ago and myself in 1981 as follows:

(i) conversion of active drug to an inactive derivative by enzyme(s) produced by the resistant cells,

(ii) modification of the drug-sensitive site,

(iii) loss of cell permeability to a drug,

(iv) synthesis of an additional drug-resistant enzyme or over-production of a drug-sensitive enzyme,

(v) increased concentration of a metabolite that antagonizes the inhibitor.

The idea that there were multidrug efflux pumps was not recognized until the early 1980s and therefore it was not on either Davis's or my list.

Above, I have discussed the events that characterized the history of antibiotics acting on murein and the classes of targets that have been exploited. To it should be added the discussion on this subject presented by Wong and Pompliano (1998).

EVOLUTION OF ANTIBIOTIC RESISTANCE WITH AND WITHOUT MANKIND'S INTERFERENCE

When life first arose, there would have been no purpose for antibiotics and the biosynthetic pathways to make them would not have developed. It has been argued (Koch, 1994) that up to the time of the Last Universal Ancestor (LUA), there would have been no stable diversity, and life was essentially a monoculture and therefore there was neither predator nor prey. Carl Woese feels that at this time there was much diversity. Self-evidently, if there was such, it was not stable. With the establishment of stable diversity it was possible for apparently competitive organisms to coexist because the competition did not lead in such circumstances to invariable extinction of one or the other form. With diversity, niches developed for organisms of a variety of sizes, species, and life strategies. Predators, prey, commensal, symbiotic, and antibiotic organisms now all could prosper. The establishment of niches (or roles for different organisms) and habitats (or different locations and environments for different organisms) is necessary for the development of complexity in an ecosystem. Antibiotic compounds would then have had a functional role and, no doubt, various organisms produced very large numbers of antibiotic chemicals. This is because intermediary metabolism develops by theme and variation of existing enzyme genes, and thus families of related natural products would have arisen.

With the introduction of natural antibiotics, selection for resistance mechanisms occurred in the target organisms. Mechanisms developed in a number of broad classes. The major categories are: (i) reduced permeability, (ii) inactivation of the antibiotic, (iii) alteration of the target within the host, (iv) development of a substitute (but resistant) biochemical pathway, (iv) and development of efflux pathways. The different types of resistance mechanism have been considered in many reviews. I will single out my own review of 1981 for two reasons. First, it listed all the classes now known, except for multidrug resistance. (The mechanism for the efflux of tetracycline was known, but the existence of mechanisms that would cause the efflux of a variety of quite different agents was not yet contemplated.) This success of the science of twenty five years ago to have found all varieties of classes now known except that of the general efflux pathway mechanisms suggests that maybe there are no unrecognized classes of resistance mechanism left in the biosphere. This is parallel to the general finding that search of the biosphere for a new antibiotic (of known class) has a low chance of success nowadays. Currently new molecular biological techniques to scan an organism's genomes and genes directly are being used to search further for antibiotics and targets for drugs using high throughput techniques. Besides members of the known classes and the targets of known antibiotics, they may find novel classes, however, they may not. Here the alternative is suggested that scientists may be able to devise and synthesize new types of antibiotics directed towards targets for which there are no natural antibiotics and no natural resistance mechanisms anywhere in the world.

SLOW, LABORIOUS EVOLUTION OF NEW PROTEINS

Much of *de novo* evolution has occurred by duplication of a gene and then modification of one copy for a new or an improved purpose. Evolution by a series of point mutations has limitations. If the new purpose is quite disparate from the original one, then further tuning and molding of the protein structure will be needed. It must be assumed that development proceeds slowly and that duplication with alternate activation and silencing occurred repeatedly. While duplication and selection were the only evolutionary procedure initially, when mechanisms for lateral transfer arose between cells, rearrangement of genetic regions and fusing with still other genes must have soon become as important as they are today. In several papers, I have considered the kinetic effects on enzyme evolution of: (i) reversibly inactivating a gene, (ii) periodic selection, and (iii) of a tandem array of genes (see Chapter 20). With such movement of DNA from cell-to-cell mediating this transfer both for the aid of the host and the pathogen, the net effect would have been the inhibition of *de novo* evolution of the host's chromosomal resistance mechanisms.

There is another quite relevant point that needs to be stressed: it appears (Koch, 1996) that the natural evolutionary mechanisms on this planet never approached finding the most efficient possible protein structures for any given purpose. This failure is demonstrated by the contemporary results with modern recombinational techniques. With these modern laboratory genetic techniques, random replacement of the nucleotides in specific parts of a gene can be used to find much more efficient proteins than exist in the biosphere. These combinations had just never been "tried" by ordinary evolutionary mechanisms lacking mankind's intelligent input. The biological implication of the results of the "selex" experiments of the last decade is that Nature was not able to explore sequence space anywhere near as completely as the modern biochemist can, although he or she is able to scan only a small, but biologically relevant, region of a genome.

LACK OF EVOLUTION OF NOVEL ANTIBIOTIC RESISTANCE AFTER MASSIVE HUMAN USE OF ANTIBIOTICS

The second point from this old review is that it asserted that no radically new (*de novo*) resistance mechanisms had developed in the then 40 years of the antibiotic era of medicine. In fact, the resistance mechanisms important in 1981 had been occasionally generated by simple genetic changes, but mostly had preexisted the antibiotic era and represented mainly lateral gene transfers from other organisms. Source organisms would include: the producing organisms of antibiotics (as in the case of aminoglycosides, discussed above) and from those organisms whose enzymatic system were inherently resistant. Considering the massive nature of the application of antibiotics and the strong selection pressures during the last half century, it appears at first to be strange that extensive *de novo* evolution of resistance did not appear. I have argued earlier (Koch, 1995) and emphasize here that this lack of evolution was because the gene transfer from other parts of the biosphere pre-empts *de novo* evolution generally that otherwise would improve resistance mechanisms.

ROLE OF TRANSFORMATION, PLASMIDS AND VIRUSES

It has been argued extensively here that insertion of exogenous DNA into living cells is not trivial and that there are no passive, spontaneous ways for DNA to enter a living cell. Although very different mechanisms for gene transfer exist in modern organisms they are all quite elaborate and complex. The transfer mechanisms need to be complex because a highly polar nucleic acid has to be passed through a hydrophobic lipid bilayer. In the modern world, transmissible plasmids and viruses can do it, some bacteria under certain conditions do have genetic mechanisms allowing them to take up naked DNA, and of course, a sperm can pass its DNA into the accepting egg by a very sophisticated process. So the argument is that, before the development of these gene-transfer mechanisms, much *de novo* creation and perfection of genetic apparatus had to take place to evolve antibiotic-resistance mechanisms inside a cell, and such ability was not passed laterally from cell to cell until much later.

After lateral transport evolved and could efficiently take place, then it would happen that some possibly poorly effective, but existing mechanism, from almost any place in the biosphere, could out compete the *de novo* development or improvement of an existing resistance mechanism in a stressed organism. The mechanism might be very efficient in the environment from which it came, but would be less efficient in the new one. This would impede *de novo* evolution generally (Koch, 1981, 1994). Although production of antibiotics in nature probably was never near the intensity of current industrial production and clinical use, mobilization of a resistance gene to a targeted microorganism by lateral transport of antibiotic resistance genes could have occurred in the distant past. However, such movement and development of resistance was not at the breakneck pace of the last 60 years.

ANTIBIOSIS AND COUNTER-MEASURES OF THE ANTIBIOTIC ERA

LACTAM ANTIBIOTICS AND VARIANTS TODAY

Looking forward from the 1940s, no one could foresee the consequences, both the great advances and the great failures that now have been observed. With the medical use of antibiotics, resistance quickly became apparent. The first was the finding by Abraham and Chain (1940) that some strains of *Escherichia coli* were already resistant, and shortly after that, Kirby (1944) found resistant strains of *Staphylococcus aureus*. These resistance mechanisms were no doubt generated by the prior existence of β-lactamases present in various places in the microbial world.

The β-lactamases are all generally related to each other and to all PBPs. This indicates that they all go back to the primordial penicillin-binding protein, PBP. There are many variants and their sequences and the constructed phylogenetic tree suggest that a large number of β-lactamase genes arose independently and were in the world's gene pool prior to man's involvement in chemotherapy.

In the late 1950s, new β-lactams were discovered, such as cephalosporin C and 6-aminopenicillianic acid. These permitted chemists to produce lactams that resisted some of the lactamases. These included the semi-synthetic isoxazolyly penicillins, such as methicillin, oxacillin, and the "first generation" cephalosporins. The use of cepalothin, cephaloridine, and cefazolin significantly decreased the resistance role of β-lactamases that was then rampant in gram-positive organisms.

EXTENDED-SPECTRUM BETA LACTAMASES (ESBL) AND CEPHALOSPORINASES

Not too surprisingly from our vantage point, antibiotics including methicillin and oxacillin were associated with the appearance of extended-spectrum β-lactamases (ESBL) and cephalosporinases. In the middle of the 1960s new plasmid encoding β-lactamases appeared in Gram-negative organisms that had been subject to cephalosporins. These enzymes have been designated SHV-1, TEM-1, and TEM-2. Of these, TEM-1 has caused particularly grave medical problems. This danger led to a concerted attack on TEM-1 by the pharmaceutical industry and resulted in a two-pronged approach. One approach was the development of β-lactams drugs containing clavulanic acid and a penicillin-derived sulfone. These were offered to the patient together with a β-lactam that blocked certain PBPs. This worked because the clavulanic acid irreversibly tied up the β-lactamase, thus protecting the β-lactam antibiotic that was then able to inhibit PBPs and block bacterial wall growth. The alternative prong was to find and use new lactams that were resistant to the β-lactamases, like the carbapenems. Later some of the other "third generation" cephalosphorins and still later the monobactam, aztreonam was developed.

The conclusion as I see it is that a completely new look for antibiotics is needed.

WHAT TO DO AND WHERE TO LOOK

Table 21.1 gives a list of the stages of murein synthesis and includes the incorporation of precursors into the stress-bearing wall and the removal of murein from the wall. The major target (due to β-lactam) is given in bold type.

Targets that one might suspect could be fruitful bear superscript numbers and I hope that at least some of these targets are susceptible to inhibition by new compounds. While it is possible that some of the methods now used by the pharmaceutical companies may generate compounds, it may be that because of the gene sequences available today, locating genes for perspective targets may be

Table 21.1 Targets for potentially novel chemotherapeutic agents acting on bacterial murein for the case of *E. coli or B. subtilis*

- Exploitation Sites
  - In cytoplasm
    - Blockade of formation of UDP-N-Acetyl muramic acid (UDP-MurNAc)
    - Blockade of formation of D-Glutamic acid (D-Glu)
    - Blockade of formation of D-Ala-D-Ala
    - Blockage of muropeptide formation
      - Linkage of L-Ala
      - Linkage of D-Glu via the γ-linkage of D-Glu
      - Linkage of Diaminopimelic acid (m-A_2pm or DAP)
      - Linkage of D-Ala-D-Ala
  - On the inside of the cytoplasmic membrane:
    - Blockage of the link of UDP-MurNAc penta-muropeptide with P-bactoprenol to yield Lipid I
    - Combination of Lipid I with UDP-GlcNAc to give Lipid II
  - Transport through the membrane[1]
  - Insertion into growing murein wall
    - Inserting disaccharide penta-muropeptide into glycan[2]
    - **The endo-transpeptidase tail-to-tail bond formation**[3]
  - Stimulation of inappropriate cleavages
    - Accelerating the critical autolysin (i.e., endo-lytic transglycosylase)[4]
    - Autolysis at inappropriate sites
    - Truncation of the muropeptides that are pointing up or down from the plane of the saccular wall[5]

[1] Getting the disaccharide penta-muropeptide through the wall must be difficult because it is hydrophilic.
[2] Inserting a disaccharide penta-muropeptide between the bactoprenol and an existing glycan chain is not trivial and could be blocked by an agent that only enters into the periplasmic space.
[3] Although β-lactams inhibit the coupling of the muropeptides, the requirements of a molecule to fit into the endopeptidase could be met by other chemically different agents that could then could inhibit the endopeptidase.
[4] Further studies of the lytic and other transglycosides could yield new antibiotics.
[5] Muropeptides pointing up or down from the plane of the murein could be targets. An agent either by inhibiting their removal or serving as points of attachment for binding of inhibitors could be an antibiotic.

more productive. Study of the steps to move precursors through the membrane and attach them to the growing murein may be a more rapid route to new classes of antibiotics (Heijenoort, 2002; Barbosa *et al.*, 2002; Holtje, 2001).

THE END GAME

The game will never be over in the chess match between bacteria, eukaryotic organisms, mankind, and medical scientists and engineers. The overall incidence of a disease depends on the number of extant individuals (at steady state the carrying capacity of the ecosystem). The interrelation of these four groups as well as their individual abilities is fundamental. But the available life styles are very critical. For example, in caveman days diseases that require a high abundance of humans would not prosper, whereas sexually transmitted diseases would prosper because of the behaviors of the host that the pathogens could exploit. Two alternative strategies of the pathogen at any time are that it may be able to grow in an alternative (and possibly abundant) host or it may be able to grow in an abiotic environment. So we can understand that in ancient Egypt, when the agriculture became adequate to enable the development a high concentration of Egyptians (and even temporarily a high concentration of Jews) it was possible that tuberculosis could became abundant. As another example, it is because of medical and industrial advances (particularly in methods of transportation) that AIDS could become abundant today. (We can debate whether increase in promiscuity is also a factor.) But we can conclude that the access to the host was critical and that the evolution and molecular biology, if evolution development were at all possible, would allow the pathogen to attack.

The challenge is for human technology to reduce or make impossible the attack by bacteria able to evolve their molecular biology rapidly. The point of this book is that, if we pose a new problem to the pathogens that would require the evolution *de novo* of a totally new protein or proteins, the time to emergence of an improved kind of pathogen will be long; longer—we can hope—than the duration of patent protection.

References

Abramson, E. P., and E. Chain. 1940. An enzyme from bacteria able to destroy penicillin. Nature **146**:837.

Atwood, K. C., L. K. Schneider, and F. J. Ryan. 1951. Periodic selection in Escherichia coli. Proc. Natl. Acad. Sci. USA **37**:146–55.

Barbosa, M. D., S. Lin, J. A. Markwalder, J. A. Mills, J. A. DeVito, C. A. Teleha, V. Garlapati, C. Liu, A. Thompson, G. L. Trainor, M. G. Kurilla, and D. L. Pompliano. 2002. Regulated expression of the *Escherichia coli lepB* gene as a tool for cellular testing of antimicrobial compounds that inhibit signal peptidase I *in vitro*. Anti Agents Chemo **46**:3549–54.

Bardy, S. L., S. Y. M. Ng, and K. F. Jarrell. 2003. Prokaryotic motility structures. Microbiol. **149:**295–304.

Barlow, M., and B. G. Hall. 2002. Phylogenetic analysis shows that OXA β-lactamase genes have been on plasmids for millions of years. J. Mol. Evol. **55**:314–21.

Benveniste, R., and J. E. Davies. 1973. Aminoglycosides antibiotic-inactivating enzymes in actinomycetes similar to those present in clinical isolates of antibiotic-resistant bacteria. Proc. Natl. Acad. Sci. USA **70**:2276–80.

Brack, A. (ed.) 1998. *The molecular origins of life*. Cambridge University Press, Cambridge.

Bush, K. 1997. The evolution of β-lactamases. *In* D. Chadwick and J. Goode (ed.), *Antibiotic resistance: origins, evolution, selection and spread.* John Wiley and Sons, Chichester, pp. 152–66.

Bush, K. 1999. β-Lactamases of increasing clinical importance. Curr. Pharm. Des. **5**:839–45.

Bush, K. 2001. New beta-lactamases in gram-negative bacteria: diversity and impact on the selection of antimicrobial therapy. Clinical Infect. Dis. **32**:1085–9.

Bush, K, and S. Mobashery. 1998.How β-lactamases have driven pharmaceutical drug discovery: from mechanistic knowledge to clinical circumvention. *In* B. P. Rosen and S. Mobashery (ed.), *Resolving the antibiotic paradox.* Kluwer Academic/Plenum Publishers, New York, pp. 71–98.

Cavalier-Smith, T. 1987. The origin of cells: a symbiosis between genes, catalyst, and membranes. Cold Spring Harb. Symp. Quant. Biol. **52**:805–24.

Cavalier-Smith, T. 2001. Obcells as proto-organisms: membrane heredity, lithophosphorylation, and the origins of the genetic code, the first cells, and photosynthesis. J. Mol. Evol. **53**:555–95.

Cavalier-Smith, T. 2002. The neomuran origin of Archaebacteria, the negibacterial root of the universal tree and the bacterial megaclassification. Int. J. Syst. Evol. Microbiol. **52**:7–76.

Chang, S., R. Mack, S. L. Miller, and S. L. Strathearn. 1983. Prebiotic organic syntheses and the origin of life. *In* J. W. Schopf *et al.*, (ed.), *Earth's earliest biosphere: its origin and evolution*. Princeton University Press, Princeton, pp. 53–92.

Charon, N. W., and S. F. Goldstein. 2002. Genetics of motility and chemotaxis of a fascinating group of bacteria: the spirochetes. Ann. Rev. Genet. **36**:47–73.

Cole, R. M., and J. J. Hahn. 1962. Cell wall replication in *Streptococcus pyogenes.* Science **135**:722.

Davies, J. E. 1997. Origins, acquisition and dissemination of antibiotic resistance determinants. Ciba Found. Symp. **207**:15–27.

Davis, B. D. 1993. The penicillin method of mutant selection. Bioessays. **15**:837–9.

De Pedro, M. A., J. C. Quintela, J.-V. Höltje, and H. Schwarz. 1997. Murein segregation in *Escherichia coli.* J. Bacteriol. **179**:2823–34.

De Pedro, M. A., H. Schwarz, and A. L. Koch. 2003. Patchiness of insertion of murein in the sidewall of *E. coli.* Microbiology **149**:1753–61.

Deamer, D. W., and J. L. Bada. 1997. The first living systems: a bioenergetic perspective. Micro. Mol. Biol. Rev. **61**:239–61.

Deamer, D. W., and G. R. Fleischaker. 1994. *Origins of life: the central concepts.* Jones and Bartlett Publishers, Boston.

Demerec, M. 1948. Origin of bacterial resistance to antibiotics. J. Bacteriol. **56**:63–74.

Dworkin, M. (ed.) *The Prokaryotes.* Springer-Verlag, Heidelberg Online.

Dykhuizen, D. E. 1998. Santa Rosalia revisited: Why are there so many species of bacteria? Antonie Van Leeuwenhoek J. Microbiol. **73**:25–33.

Eigen, M., and P. Schuster. 1977. The hypercycle. A principle of natural self-organization. Part A: Emergence of the hypercycle. Naturwissenschaften **64**:541–65.

Eigen, M., and P. Schuster. 1982. Stages of emerging life—five principles of early organization. J. Mol. Evol. **19**:47–61.

Enright, M. C., D. A. Robinson, R. Gaynor, E. J. Feil, H. Grundmann, and B. W. Spratt. 2002. The evolutionary history of methicillin-resistant *Staphylococcus aureus* (MRSA). Proc. Natl. Acad. Sci. USA **99**: 7687–92.

Gause, G. F. 1971. *The struggle for existence.* Dover Publications, New York

Ghuysen, J. M., and R. Hakenbeck (ed.). 1994. *Bacterial cell wall.* Elsevier, Amsterdam.

Gilad, R., A. Porat, and S. Trachtenberg. 2003. Motility modes of *Spiroplasma melliferum* BC3: a helical, wall-less bacterium driven by a linear motor. Mol. Microbiol. **47**:657–69.

Griffith, A. A. 1920. The phenomena of rupture and flow in solids. Phil. Trans. R. Soc. A **221**:63–98.

Gupta, R. S. 2001. The branching order and phylogenetic placement of species from completed bacterial genomes, based on conserved indels found in various proteins. Int. Microbiol. **4**:187–202.

Gupta, R. S. 2002. Phylogeny of bacteria, are we now close to understanding it? ASM News **68**:284–91.

Gupta, R. S., and E. Griffiths. 2002. Critical issues in bacterial phylogeny. Theor. Popul. Biol. **61**:423–34.

Harz, H., K. Burgdorf, and J.-V. Höltje. 1990. Isolation and separation of the glycan strands from murein of *Escherichia coli* by reversed-phase high-performance liquid chromatography. Anal. Biochem. **190**:120–8.

Hiramatsu K., L. Cui, M. Kuroda, and T. Ito. 2001.The emergence and evolution of methicillin-resistant *Staphylococcus aureus*. Trends Microbiol. **9**:486–93.

Hiramatsu, K. 1995. Molecular evolution of MRSA. Microbiol. Immunol. **39**:531–43.

Höltje, J.-V. 1993. "Three for One"—A simple growth mechanism that guarantees a precise copy of the thin, rod-shaped sacculus *of E. coli. In* M. A. de Pedro, J.-V. Höltje, and W. Löffëlhardt, (ed.), Bacterial growth analysis. Metabolism and structure of the bacterial sacculus. Plenum Press, New York, pp. 419–26.

Höltje, J-.V. 2001. The alternatives to penicillins. Nat. Med. **7**:1100–1.

Höltje, J.-V., and U. Schwarz. 1985. Biosynthesis and growth of the murein cell. *In* N. Nanninga (ed.), *Molecular cytology of Escherichia coli.* Academic Press, New York, pp. 161–97

Ishidate, K., A. Ursinus, J.-V. Höltje, and L. Rothfield. 1998. Analysis of the length distribution of murein glycan strands in ftsZ and ftsI mutants of *E. coli.* FEMS Microbiol. Lett. **168**:71–5.

Jacob, F., S. Brenner, and F. Cuzins. 1963. On the regulation of DNA replication in bacteria. Cold Spring Harb. Symp. Quant. Biol. **2:**239–347.

Jollès, P. 1996. *Lysozymes: model enzymes in biochemistry and biology.* Birkhäuser Verlag, Basel

Jones, L. J., R. Carballido-Lopez, and J. Errington. 2001. Control of cell shape in bacteria: helical, actin-like filaments in *Bacillus subtilis.* Cell **104**:913–22.

Kirby, W. M. M. 1944. Extraction of a potent penicillin inactivator from penicillin-resistant staphylococci. Science **99**:452–3.

Koch, A. L. 1972. Enzyme evolution: the importance of untranslatable intermediates. Genetics **72**:297–316.

Koch, A. L. 1974. The pertinence of the periodic selection phenomenon to procaryote evolution. Genetics **77**:127–42.

Koch, A. L. 1979. Microbial growth in low concentrations of nutrients. *In* M. Shilo (ed.), *Strategies in microbial life in extreme environments.* Dahlem Konferenzen-1978, Berlin, pp. 261–79.

Koch, A. L. 1980. Selection and recombination in populations containing tandem multiplet genes. J .Mol. Evol. **14**:273–85.

Koch, A. L. 1981. Evolution of antibiotic resistance gene function. Microbiol. Rev. **45**:355–78.

Koch, A. L. 1983. The surface stress theory of microbial morphogenesis. Adv. Microbiol. Physiol. **24**:301–66.

Koch, A. L. 1984a. Evolution vs. the number of gene copies per primitive cell. J. Mol. Evol. **20**:71–6.

Koch, A. L. 1984b. Shrinkage of growing *Escherichia coli* cells through osmotic challenge. J. Bacteriol. **159**:914–24.

Koch, A. L. 1985. Primeval cells: possible energy-generating and cell-division mechanisms. J. Mol. Evol. **21**:270–7.

Koch, A. L. 1988. Biophysics of bacterial wall viewed as a stress-bearing fabric. Microbiol. Rev. **52**:337–53.

Koch, A. L. 1990a. Growth and form of the bacterial cell wall. Am. Sci. **78**:327–41.

Koch, A. L. 1990b. The relative rotation of the ends of *Bacillus subtilis*. Arch. Microbiol. **153**:569–73.

Koch, A. L. 1990c. The surface stress theory for the case of *E. coli*: the paradoxes of gram-negative growth. Res. Microbiol. **141**:119–30.

Koch, A. L. 1991. The wall of bacteria serves the role that mechano-proteins do in eukaryotes. FEMS Microbiol. Rev. **88**:15–26.

Koch, A. L. 1993. Microbial genetic responses to extreme challenges. J. Theor. Biol. **160**:1–21.

Koch, A. L. 1994. Development and diversification of the Last Universal Ancestor. J. Theor. Biol. **168**:269–80.

Koch, A. L. 1995. Origin of intracellular and intercellular pathogens. Quart. Rev. Biol. **70**:423–37.

Koch, A. L. 2000a. Simulation of the conformation of the murein fabric: I. The oligoglycan, penta-muropeptide, and crosslinked nona-muropeptide. Arch. Microbiol. **174**:429–39

Koch, A. L. 2000b. Penicillin binding proteins, β-lactams, and lactamases: Offensives, attacks and defensive counter measures. Crit. Rev. Microbiol. **26**:1–35.

Koch, A. L. 2000c. The bacterial way for safe enlargement and division. App. Env. Micro. **66**:3657–63.

Koch, A. L. 2001. *Bacterial growth and form.* 2nd edn, Kluwer Academic Publishers, Dordrecht, The Netherlands.

Koch, A. L. 2002. Why are rod-shaped bacteria rod shaped? Trends Microbiol. **10**:452–5.

Koch, A. L. 2003a. Were Gram-positive rods the first bacteria? Trends Microbiol. **11**:166–10.

Koch, A. L. 2003b. Cell wall-deficient (CWD) bacterial pathogens: could amyotrophic lateral sclerosis (ALS) be due to one? Crit. Rev. Microbiol. **29**:215–21.

Koch, A. L., and I. D. J. Burdett. 1984. The variable-T model for Gram-negative morphology. J. Genet. Microbiol. **130**:2325–38.

Koch, A. L., M. L. Higgins, and R. J. Doyle. 1981. Surface tension-like forces determine bacterial shapes: *Streptococcus faecium*. J. Genet. Microbiol. **123**:151–61.

Koch, A. L., M. L. Higgins, and R. J. Doyle. 1982. The role of surface stress in the morphology of microbes. J. Genet. Microbiol. **128**:927–45.

Koch, A. L., and M. L. Higgins. 1984. Control of wall band splitting in *Streptococcus faecalis* ATCC 9790. J. Genet. Microbiol. **130**:735–45.

Koch, A. L., S. L. Lane, J. Miller, and D. Nickens. 1987. Contraction of filaments of *Escherichia coli* after disruption of the cell membrane by detergent. J. Bacteriol. **166**:1979–84.

Koch, A. L., H. L. T. Mobley, R. J. Doyle, and U. N. Streips. 1981. The coupling of wall growth and chromosome replication in Gram-positive rods. FEMS Microbiol. Lett. **2**:201–8.

Koch, A. L., and M. F. S. Pinette. 1987. Nephelometric determination of osmotic pressure in growing gram-negative bacteria. J. Bacteriol. **169**:3654–63.

Koch, A. L., and S. Silver. 2005. The first cell. FEMS Microbiol Rev. Advances in Microbiol Physiology. **50**:227–259.

Koch, A. L., and S. W. Woeste. 1992. The elasticity of the sacculus of *Escherichia coli*. J. Bacteriol. **17:**4811–9.

Kraft, A. R., J. Prabhu, A. Ursinus, and J.-V. Höltje. 1999. Interference with murein turnover has no effect on growth but reduces beta-lactamase induction in *Escherichia coli*. J. Bacteriol. **181**:7192–8.

Labischinski, H., G. Barnickel, H. Bradaczek, and P. Giesbrecht. 1979. On the secondary and tertiary structure of murein. Eur. J. Biochem. **95**:147–55.

Labischinski, H., E. W. Goodell, A. Goodell, and M. L. Hochberg. 1991. Direct proof of a "more-than-single-layered" peptidoglycan architecture of *Escherichia coli* W7: a neutron small-angle scattering study. J. Bacteriol. **173**:751–6.

Labischinski, H., G. Barnickel, D. Naumann, and P. Keller. 1983. Conformational and topical aspects of the three-dimensional architecture of bacterial peptidoglycan. Ann. Microbiol. (Institut Pasteur) A **136**:45–50.

Lee, A., and J. L. O'Rourke. 1993. Ultrastructure of Helicobacter organisms and possible relevance for pathogenesis *In* C. S. Goodwin and B. W. Worsley (ed.), *Helicobacter pylori*—biology and clinical practice. CRC Press, Boca Raton, FL, pp. 15–35.

Leps, B., H. Labischinski, G. Barnickel, H. Bradaczek, and P. Giesbrecht. 1984. A new proposal for the primary and secondary structure of the glycan moiety of pseudomurein. Conformational energy calculations on the glycan strands with talosaminuronic acid in 1C conformation and comparison with murein. Eur. J. Biochem. **144**:279–86.

Li, C., M., A. Motaleb, M. Sal1, S. F. Goldstein, and N. W. Charon. 2000. Spirochete motility J. Mol. Micro. Biotechnol. **2**:345–54.

Li, C., R. G. Bakker, M. A. Motaleb, M. L. Sartakova, F. C. Cabello, and N. W. Charon. 2004. Asymmetrical flagellar rotation in *Borrelia burgdorferi* nonchemotactic mutants. Proc. Natl. Acad. Sci. USA. **99**:6169–74.

Luria, S., and M. Delbrück. 1943. Mutations of bacteria from virus sensitivity to virus resistance. Genetics. **28**:491–511.

Macnab, R. M. 1999. The bacterial flagellum: reversible rotary propeller and type III export apparatus. J. Bacteriol. **181**:7149–53.

Mark, H. 1943. Elasticity and strength. *In* E. Ott (ed.), *Cellulose and its derivatives.* Interscience, New York, pp. 990–1052

Massova, I., and S. Mobashery, 1997. Molecular bases for interactions between β-lactams antibiotics and β-lactamases. Acc. Chem. Res. **30**:162–8.

Massova, I., and S. Mobashery 1998. Kinship and diversification of bacterial penicillin-binding proteins and β-lactamases. Antimicrob. Agents Chemo. **77**:1–17.

Massova, I., and S. Mobashery. 1999. Structural and mechanistic aspects of β-lactamases and penicillin-binding proteins. Curr. Pharm. Des. **5**:929–37.

Mattman, L. H. 1974. *Cell wall deficient forms*. CRC Press, Cleveland

Mattman, L. H. 2001. *Cell deficient forms: stealth pathogens*. 3rd edn CRC Press, Boca Raton, FL

Mazel, D. and J. Davies. 1999. Antibiotic resistance in microbes. Cell. Mol. Life Sci. **56**:742–54.

Medeiros, A. A. 1997. Evolution and dissemination of β-lactamases, accelerated by generations of β-lactam antibiotics. Clinical Infect. Dis. (Supplement 1) **24** :S19–45.

Mendelson, N. H. 1976. Helical growth of *Bacillus subtilis*: a new model for cell growth. Proc. Natl. Acad. Sci. USA. **73**:1740–4.

Menge, H., M. Gregor, G. N. J. Tytgat, B. J. Marshall, and C. A. M. McNulty. 1990. *Helicobacter pylori* 1990 Springer Verlag, Berlin.

Merad, T., A. R. Archibald, I. C. Hancock, C. R. Harwood, and J. A. Hobot. 1989. Cell wall assembly in *Bacillus subtilis*: visualisation of old and new material by electron microscopic examination of samples selectively stained for teichoic acid and teichuronic acid. J. Gen. Microbiol. **135**:645–55.

Miller, S. L., and L. E. Orgel. 1973. *The origins of life on earth*. Prentice-Hall, New York.

Mobley, M. L. T., G. L. Mendz, and S. L. Hazell. 2001. *Helicobacter pylori*. ASM Press, Washington.DC

Morin, R. B., and M. Gorman. 1982. The biosynthesis of β–lactam antibiotics. *Biochemistry*, vol 3. Academic Press, London.

Norris, V. 1992. Phospholipid domains determine the spatial organization of the *Escherichia coli* cell cycle: The membrane tectonics model. J. Theor. Biol. **154**:91–107.

Obermann, W., and J.-V. Höltje. 1994. Alterations of murein structure and of penicillin-binding proteins in minicells from *Escherichia coli*. Microbiol. **140**:79–87.

Olsen, G. J. 2001. The history of life. Nat. Genet. **28**:197–8.

On, S. L., A. Lee, J. L. O'Rourke, F. E. Dewhirst, B. J. Paster, J. G. Fox, and P. A. R. Vandamme. 2002. Genus *Helicobacter. In* G. M. Garrity, and D. J. Brenner (ed.), *Bergey's manual of systematic bacteriology* Bergey's Manual Trust, New York, NY.

Osserman, E. F., R. E. Canfield, and S. Beychok. 1972. *Lysozyme*. Academic Press, New York

Pinette, M. F. S., and A. L. Koch. 1988b. Turgor pressure responses of a gram-negative bacterium to antibiotic treatment measured by collapse of gas vesicles. J. Bacteriol. **170**:1129–36.

Previc, E. D. 1970. Biochemical determination of bacterial morphology. J. Theor. Biol. **24**:471–497.

Queener, S. W., and N. Neuss. 1982. Chemistry and biology of β-lactam antibiotics *In* R. B. Morin and M. Gorman (ed.), *Biochemistry, vol. 3 The biosynthesis of β–lactam antibiotics*. Academic Press, London, pp. 2–81

Rogers, H. J., H. R. Perkins, and J. B. Ward. 1980. *Microbial cell walls and membranes.* Chapman and Hall, London.

Schliefer, K. H., and O. Kandler. 1972. Peptidoglycan types of bacterial cell wall and their taxonomic implications. Bacteriol. Rev. **36**:407–77.

Seifert, J. L., and G. E. Fox. 1998. Phylogenetic mapping of the bacteria morphology. Microbiol. **144**:2803–8.

Tamames, J., M. Gonzalez-Moreno, J. Mingorance, A. Valencia, and M. Vincente 2001. Bringing gene order into bacterial shape. Trends Genet. **17**:124–6.

Tipper, D. J., and A. Wright. 1979. The structure and biosynthesis of bacterial cell walls. *In* J. R. Sokatch and L. N. Ornston (ed.), *The bacteria*, vol. 7. Academic Press, London, pp. 291–426.

Thompson, G. L., M. G. Trainor, M. G. Kurilla, and D. L. Pompliano. 2002. Regulated expression of the *Escherichia coli* lepB gene as a tool for cellular testing of antimicrobial compounds that inhibit signal peptidase I *in vitro*. Anti Agents Chemo **46**:3549–54.

Trachtenberg, S. 1998. Mollicutes—Wall-less bacteria with internal cytoskeletons. J. Struct. Biol. **124**:244–56.

Trachtenberg, S., and R. Gilad. 2001. A bacterial linear motor: cellular and molecular organization of the contractile cytoskeleton of the helical bacterium *Spiroplasma melliferum* BC3. Mol. Microbiol. **41**:827–48.

van Heijenoort, J. 2001. Formation of the glycan chains in the synthesis of bacterial peptidoglycan. Glycobiology **11**:25R–36R.

Verwer, R. W. H., N. Nanninga, W. Keck., and U. Schwarz. 1978. Arrangements of glycan chains in the sacculus of *Escherichia coli.* J. Bacteriol. **136**:723–9.

Wächtershäuser, G. 1988a. Before enzymes and templates: theory of surface metabolism. Microbiol. Rev. **52**:452–84.

Wächtershäuser, G. 1988b. Pyrite formation, the first energy source for life: a hypothesis. Syst. Appl. Microbiol. Biophys. **10**:207–10.

Wächtershäuser, G. 1990. Evolution of the first metabolic cycles. Proc. Natl. Acad. Sci. USA **87**:200–4.

Wächtershäuser, G. 1993. Ground work for evolutionary biochemistry: the iron-sulphur world. Prog. Biophys. Mol. Biol. **58**:85–201.

Wächtershäuser, G. 1994. Life in a ligand sphere. Proc. Natl. Acad. Sci. USA **91**:283–4287.

Weidel, W., and H. Pelzer. 1964. Bag-shaped macromolecules?a new outlook on bacterial cell walls. Adv. Enzymol. **26**:193–232.

Woese, C. R. 1979. A proposal concerning the origin of life on the planet earth. J. Mol. Evol. **13**:95–101.

Woese, C. R. 1987. Bacterial evolution. Micro. Rev. **51**:221–71.

Woese, C. R. 1998. The universal ancestor. Proc. Natl. Acad. Sci. USA. **98**:6854–9.

Woese, C. R. 2000. Interpreting the universal phylogenetic tree. Proc. Natl. Acad. Sci. USA **97**: 8392–6.

Woese, C. R., O. Kandler, and M. L. Wheelis. 1990. Towards a natural system of organisms: proposal for the domains Archaea, Bacteria, and Eucarya, Proc. Natl. Acad. Sci. USA **87**:4576–9.

Woldringh, C. L., N. B. Grover, R. F. Rosenberger, and A. Zaritsky. 1980. Dimensional rearragement of rod-shaped bacteria following a nutritional shift-.up II Experiments with Escherichia coli B/r. J. Theor. Biol. **86**:441–54.

Woldringh, C. L., and N. Nanninga. 1985. Structure of nucleoid and cytoplasm in the intact cell. *In* N. Nanninga (ed.), *Molecular cytology, Escherichia coli.* Academic Press, New York, pp. 161–97.

Wong, K. K., and D. L. Pompliano. 1998. Peptoglycan biosynthesis: unexploited antibacterial targets within a familiar pathway. *In* B. P. Rosen, and S. Mobashery (ed.), *Resolving the antibiotic paradox.* Kluwer Academic/ Plenum Publishers, New York, pp. 197–217.

Young, K. D. 2003. Bacterial shape. Mol. Microbiol. **49**:571–80.